Advances in Integrated Energy Systems Design, Control and Optimization

Special Issue Editors

Josep M. Guerrero

Amjad Anvari-Moghaddam

MDPI • Basel • Beijing • Wuhan • Barcelona • Belgrade

MDPI

Special Issue Editors
Josep M. Guerrero
Aalborg University
Denmark

Amjad Anvari-Moghaddam
Aalborg University
Denmark

Editorial Office
MDPI AG
St. Alban-Anlage 66
Basel, Switzerland

This edition is a reprint of the Special Issue published online in the open access journal *Applied Sciences* (ISSN 2076-3417) from 2016–2017 (available at: http://www.mdpi.com/journal/applsci/special_issues/integrated_energy).

For citation purposes, cite each article independently as indicated on the article page online and as indicated below:

Author 1, Author 2. Article title. *Journal Name*. **Year**. Article number/page range.

First Edition 2017

ISBN 978-3-03842-490-1 (Pbk)
ISBN 978-3-03842-491-8 (PDF)

Table of Contents

About the Special Issue Editors

Josep M. Guerrero received the B.S. degree in telecommunications engineering, the M.S. degree in electronics engineering, and the Ph.D. degree in power electronics from the Technical University of Catalonia, Barcelona, in 1997, 2000 and 2003, respectively. Since 2011, he has been a Full Professor with the Department of Energy Technology, Aalborg University, Denmark, where he is responsible for the Microgrid Research Program. From 2012 he is a guest Professor at the Chinese Academy of Science and the Nanjing University of Aeronautics and Astronautics; from 2014 he is chair Professor in Shandong University; from 2015 he is a distinguished guest Professor in Hunan University; and from 2016 he is a visiting professor fellow at Aston University, UK, and a guest Professor at the Nanjing University of Posts and Telecommunications. His research interests is oriented to different microgrid aspects, including power electronics, energy-related systems operation, control and optimization.

Amjad Anvari-Moghaddam received the Ph.D. degree (Hon.) from University of Tehran, Tehran, Iran, in 2015 in Power Systems Engineering. Currently, he is a Postdoctoral Fellow at the Department of Energy Technology, Aalborg University. His research interests include smart microgrids, optimal control and management, and integrated energy systems. He has authored and co-authored 6 book chapters and more than 60 technical papers in energy systems operation and control. Dr. Anvari-Moghaddam is the Guest Editor of the IEEE TRANSACTIONS ON INDUSTRIAL INFORMATICS special issue: "Next Generation Intelligent Maritime Grids", the IEEE INTERNET OF THINGS JOURNAL, special issue: "IoT-enabled Smart Energy Systems", and the Editorial Board Member for SCIREA Journal of Electrical Engineering. He is also a Senior Member of IEEE, member of Technical Committee (TC) of Renewable Energy Systems-IEEE IES, TC Member of IES Resilience and Security for Industrial Applications-IEEE PES, TC member of IEEE Working Group P2004 (HIL Simulation and Testing), and IEEE Young Professionals.

Preface to "Advances in Integrated Energy Systems Design, Control and Optimization"

1. Introduction

In the face of climate change and resource scarcity, energy supply systems are on the verge of a major transformation, which mainly includes the introduction of new components and their integration into the existing infrastructures, new network configurations and reliable topologies, optimal design and novel operation schemes, and new incentives and business models. This revolution is affecting the current paradigm and demanding that energy systems be integrated into multi-carrier energy hubs [1]. It is greatly increasing the interactions between today's energy systems at various scales (ranging from the multinational, national, community scales down to the building level) and future intelligent energy systems, which are able to incorporate an increasing amount of often fluctuating, renewable energy sources (RESs). It also increases the need for the integration of energy storage options into the energy mix, not only to reduce the need for increased peak generation capacity, but also to enhance grid reliability and support higher penetration of RESs [2]. Moreover, this transformation is accommodating active participation of end-users as responsive prosumers at different scales, which in turn helps to reduce energy costs to all consumers, increase reliability of service and mitigate carbon footprints. However, this plan of action necessitates regulatory frameworks, strategic incentives and business models for efficient deployment.

2. Energy Systems Design, Control and Optimization

To cover the above-mentioned promising and dynamic areas of research and development, this Special Issue was launched to allow the gathering of contributions in design, control and optimization of integrated energy systems. In total, 23 papers were submitted to this Special Issue, nine of which were selected for publication which denotes an acceptance rate of 39%. The accepted articles in this Special Issue cover a variety of topics, ranging from operation and control of small-scale electrical networks to the complex design and planning of energy systems.

In the first paper, a novel control scheme is proposed by Z. Zhu, J. Sun, and G. Qi for a frequency-controlled power grid not only to improve the frequency regulation of the power grid as well as the input-to-state stability, but also to minimize the communication cost within the study system [3]. The second paper, authored by J.-W. Choi and M.-K. Kim, studies the voltage stability of a renewable-based power system (mainly driven by wind turbines) using Monte Carlo simulations (MCS) and probabilistic security-constrained optimal power flow techniques [4]. In this paper, it is also demonstrated that as the wind energy penetration into a grid environment increases, the system voltage stability is more affected by the wind turbines due to the stochastic wind behavior. As a complement to the previous study, M. Vahedipour-Dahraie, H. Rashidizaheh-Kermani, H. Najafi, A. Anvari-Moghaddam, and J. Guerrero show how optimal scheduling and dispatch of electric vehicles (EVs) could enhance the system performance in terms of stability, considering high penetration of renewables in a typical network [5]. This could also be a good solution to the frequency instability (or weak stability) problem in islanded microgrids where there is low inertia for frequency compensation. The fourth paper in this Special Issue studies the important role of end-use consumers in optimal energy systems scheduling [6]. The work, done by M. Vahedipour-Dahraie, H. Najafi, A. Anvari-Moghaddam, and J. Guerrero, demonstrates the positive effect of various time-based rate (TBR) demand response (DR) programs on stochastic day-ahead energy and reserve scheduling. This is deemed to be a new trend in energy systems optimization where consumers change their consumption behavior in response to the changes in energy market prices or market incentives. Focusing on energy-related production and consumption units management, the fifth paper authored by J. Wang, K. Fang, J. Dai, Y. Yang, and Y. Zhou reflects on co-optimal distributed generation and load management considering task continuity constraints [7]. Moreover, this paper shows how energy management solutions can effectively be integrated into industrial applications to accurately perceive and access users' needs in an economic way. The focus of the next paper is to optimally size and allocate energy storage systems (ESSs) in an integrated energy system in a cost-effective and emission-aware fashion [8]. In this work, H. Lan, H. Yin, S. Wen, Y.-Y. Hong, D. Yu, and L. Zhang perform different case studies to clearly demonstrate that optimal battery sitting and sizing could ensure secure and economic integration of wind turbines into a power system to minimize the total operation cost and improve voltage profiles. As a real-world example, K. Pambour, B. Cakir Erdener, R. Bolado-Lavin, and G. Dijkema propose a practical simulation framework for analyzing security of supply in integrated gas and electric power systems [9]. This work, which is developed within the framework of the European Program for Critical Infrastructure Protection (EPCIP) of the European Commission, clearly paves the way for close collaboration and coordination between gas and power transmission system operators (TSOs) from an integrated energy system perspective.

Resource management in energy systems under faulty conditions is another research challenge that needs to be addressed suitably. In this Special Issue, B. Goo, S. Jung, and J. Hur tackle this timely topic by proposing a fast restoration procedure for power systems affected by blackouts [10]. They initially outline an optimal selection mechanism for black start units using generator characteristic data and advanced algorithms considering minimum restoration time as an objective. Afterwards, they verify the effectiveness and applicability of the proposed method by an empirical system in the eastern regions of South Korea. Last but not least, U. Tamrakar, D. Shrestha, M. Maharjan, B. Bhattarai, T. Hansen, and R. Tonkoski review the recent literature on the control of modern power systems (which are on the verge of transition from synchronous machine-based systems towards inverter-dominated systems) and describe the current state-of-the-art in such virtual inertia systems under high renewable energy penetration [11]. They also suggest potential research directions and challenges in this subject area.

3. Energy Systems of the Future

The trend in energy systems integration for sustainable development has been increasing over the past few years and becoming the point of interest for many researchers and scientists worldwide. As more and more conventional and centralized energy sources that used to ensure system stability are removed from the generation mix or shut down, distributed generation (DG) units together with energy storage options may form micro energy grids (MEG) and provide a solution. Such MEGs have a variety of micro-sources (both in the form of conventional and renewable sources) that are closely networked with one another and could play the role of virtual control units over larger areas. These integrated energy systems also have the potential to meet the high demands both economically and reliably.

In the future, we will have to integrate energy-related devices and interfaces (such as inverter-interfaced distributed energy sources) more intensely into energy systems not only to ensure secure optimal operation and control in normal mode, but also to maintain stability when a fault occurs. The next-generation virtual control units are also standardized and more affordable because they are less specific in their demands.

Josep M. Guerrero and Amjad Anvari-Moghaddam
Special Issue Editors

Acknowledgments: We would like to take this opportunity to thank all the authors for their great contributions in this Special Issue and the esteemed reviewers for their time spent on reviewing manuscripts and their valuable comments helping us to improve the articles. We wish also to place on record our appreciation to the dedicated editorial team of *Applied Sciences,* and special thanks to Natalie Sun, Senior Assistant Editor and Michelle Zhou, Assistant Managing Editor for their tremendous support and dedication.

Conflicts of Interest: The authors declare no conflict of interest.

References

1. Javadi, M.S.; Anvari-moghadam, A.; Guerrero, J.M. Optimal Scheduling of a Multi-Carrier Energy Hub Supplemented By Battery Energy Storage Systems. In Proceedings of the 17th annual conference of the International Conference on Environmental and Electrical Engineering (EEEIC 2017), Milan, Italy, 6–9 June 2017; pp. 1–6.

2. Mokhtari, G.; Anvari-Moghaddam, A.; Nourbakhsh, G. Distributed Control and Management of Renewable Electric Energy Resources for Future Grid Requirements. In *Energy Management of Distributed Generation Systems*, 1st ed.; Lucian, M., Ed.; InTech: Rijeka, Croatia, 2016; pp. 1–24.

3. Zhu, Z.; Sun, J.; Qi, G. Frequency Regulation of Power Systems with Self-Triggered Control under the Consideration of Communication Costs. *Appl. Sci.* **2017**, *7*, 688.

4. Choi, J.-W.; Kim, M.-K. Multi-Objective Optimization of Voltage-Stability Based on Congestion Management for Integrating Wind Power into the Electricity Market. *Appl. Sci.* **2017**, *7*, 573.

5. Vahedipour-Dahraie, M.; Rashidizaheh-Kermani, H.; Najafi, H.; Anvari-Moghaddam, A.; Guerrero, J. Coordination of EVs Participation for Load Frequency Control in Isolated Microgrids. *Appl. Sci.* **2017**, *7*, 539.

6. Vahedipour-Dahraie, M.; Najafi, H.; Anvari-Moghaddam, A.; Guerrero, J. Study of the Effect of Time-Based Rate Demand Response Programs on Stochastic Day-Ahead Energy and Reserve Scheduling in Islanded Residential Microgrids. *Appl. Sci.* **2017**, *7*, 378.

7. Wang, J.; Fang, K.; Dai, J.; Yang, Y.; Zhou, Y. Optimal Scheduling of Industrial Task-Continuous Load Management for Smart Power Utilization. *Appl. Sci.* **2017**, *7*, 281.

8. Lan, H.; Yin, H.; Wen, S.; Hong, Y.-Y.; Yu, D.; Zhang, L. Electrical Energy Forecasting and Optimal Allocation of ESS in a Hybrid Wind-Diesel Power System. *Appl. Sci.* **2017**, *7*, 155.

9. Pambour, K.; Cakir Erdener, B.; Bolado-Lavin, R.; Dijkema, G. Development of a Simulation Framework for Analyzing Security of Supply in Integrated Gas and Electric Power Systems. *Appl. Sci.* **2017**, *7*, 47.

10. Goo, B.; Jung, S.; Hur, J. Development of a Sequential Restoration Strategy Based on the Enhanced Dijkstra Algorithm for Korean Power Systems. *Appl. Sci.* **2016**, *6*, 435.

11. Tamrakar, U.; Shrestha, D.; Maharjan, M.; Bhattarai, B.; Hansen, T.; Tonkoski, R. Virtual Inertia: Current Trends and Future Directions. *Appl. Sci.* **2017**, *7*, 654.

applied
sciences

MDPI

Article

Frequency Regulation of Power Systems with Self-Triggered Control under the Consideration of Communication Costs

Zhiqin Zhu [1,2], **Jian Sun** [2,*], **Guanqiu Qi** [3] (iD), **Yi Chai** [4] **and Yinong Chen** [3]

[1]   College of Automation, Chongqing University of Posts and Telecommunications, Chongqing 400065, China; zhuzq@cqupt.edu.cn
[2]   College of Electronic and Information Engineering, Southwest University, Chongqing 400715, China
[3]   School of Computing, Informatics, and Decision Systems Engineering, Arizona State University, Tempe, AZ 85287, USA; guanqiuq@asu.edu (G.Q.); yinong@asu.edu (Y.C.)
[4]   State Key Laboratory of Power Transmission Equipment and System Security and New Technology, Chongqing University, Chongqing 400044, China; chaiyi@cqu.edu.cn
[*]   Correspondence: cq_jsun@163.com

Academic Editor: Johannes Kiefer
Received: 9 April 2017; Accepted: 30 June 2017; Published: 4 July 2017

Abstract: In control systems of power grids, conveying observations to controllers and obtaining control outputs depend greatly on communication and computation resources. Particularly for large-scale systems, the costs of computation and communication (cyber costs) should not be neglected. This paper proposes a self-triggered frequency control system for a power grid to reduce communication costs. An equation for obtaining the triggering time is derived, and an approximation method is proposed to reduce the computation cost of triggering time. In addition, the communication cost of frequency triggering is measured quantitatively and proportionally. The defined cost function considers both physical cost (electricity transmission cost) and communication cost (control signal transmission cost). The upper bound of cost is estimated. According to the estimated upper bound of cost, parameters of the controller are investigated by using the proposed optimization algorithm to guarantee the high performance of the system. Finally, the proposed self-triggered power system is simulated to verify its efficiency and effectiveness.

Keywords: frequency regulation; self-triggered control; input to state stable; communication cost

1. Introduction

Control systems have been widely used in power grids, and large-scale control systems continually increase the proportion of usage. Due to limited computation and communication resources, the traditional electric power network seriously affects the performance of large-scale control systems. The public shared networks for control signal communication, such as the Internet, have more powerful computation and communication resources which are used to increase the efficiency of current large-scale control systems. The scheduling of computation capacity and communication bandwidth is the key to efficiently utilizing public shared networks. Event-triggered control (ETC) systems are used to reduce the costs of computation and energy resources when the electricity network is in a steady state. They avoid the limited bandwidth of the communication network occupied by redundant sampled and fused signals [1–6]. The self-triggered control (STC) method [7,8] as one of the control methods can effectively reduce communication costs and power costs of sensor monitoring in the control system. An STC method is applicable to solve the optimization of scheduling computation and communication resources, especially in networked control systems (NCSs). The control task is triggered when STC is at the pre-computed updating time.

Since more and more systems have become increasingly networked, wireless, and spatially distributed, event-based systems are proposed to adopt a model of calls for resources only if necessary, and to utilize communication bandwidth, computation capability, and energy budget efficiently [9–12]. The proposed control-based model allowed each control task to trigger itself to achieve the optimization of computing resources and control performance. The execution time of next instance was scheduled by the executing instance. As a function of the utilization factor and control performance, the next instance execution point was dynamically obtained in time [13–17]. The dynamic selection of an appropriate threshold was investigated for basic send-on-delta (SoD) sampling strategies. The error reduction principle was formulated and proved to reduce the signal tracking-error in an available transmission rate [18,19]. A new event-based control strategy was proposed and applied to differential wheeled robots. Compared with the classical discrete–time strategy, the proposed event-based control strategy not only reached the same accuracy, but also obtained a higher efficiency in communication resource usage [20–23].

In power system frequency regulation research, Shashi Kant Pandey [24] used linear matrix inequalities (LMI) with parameters tuned by particle swarm optimization (PSO). Praghnesh Bhatt [25] analyzed the dynamic participation of doubly-fed induction generators and coordinated control for frequency regulation of an interconnected two-area power system in a restructured competitive electricity market. Soumya R. Mohanty [26] presented a study on frequency regulation in an isolated hybrid distributed generation (DG) system with the robust H-infinite loop shaping controller. Although much frequency regulation research has been done in order to achieve better physical performances, cyber costs also need to be taken into account. Considering both physical performance and the cyber cost of power system, event-driven schemes are usually used. Dai [27] proposed a methodology for real-time prediction that required event-driven load shedding (ELS) against severe contingency events. Jun [28] presented a novel emergency damping control (EDC) to suppress inter-area oscillations occurring as anticipated low-probability cases in power system operations. The proposed EDC combined an event-driven scheme and a response-based control strategy. Yan [29] elaborated a new approach based on parallel-differential evolution (P-DE) to efficiently and globally optimize ELS against voltage collapse.

However, continuously monitoring plants using event-driven schemes takes many cyber resources. In contrast with the EDC approach, STC does not generally require dedicated hardware to continuously monitor the plant state and check the defined stability conditions [30–33]. Therefore, STC can be considered in power grid frequency regulation to reduce communication costs and make the utilization of communication resources more efficient.

In this paper, we propose a novel self-triggered control scheme employed in a frequency-controlled power grid. A power grid consists of many subsystems that interact with each other through communication networks and power flow. The proposed self-triggered controller calculates the triggering period with each state point to ensure the system's exponential stability, input-to-state stability, and low communication cost. In the proposed model, the triggering interval is a function of system state and triggering rate is proportional to communication cost. This paper has three main contributions as follows:

- A self-triggered control scheme is applied to the frequency regulation of the power grid;
- An online optimization method is used to extend the triggering period for reducing communication cost; and
- Communication cost and parameters of control system for power grid are estimated and optimized, so that the cost of control system can be guaranteed under a required level.

This paper is organized as follows. Section 2 introduces the model of the power system and the basic concept used in this paper. Section 3 presents the proposed self-triggered control scheme in which the exponential stability theory with varying sampling rate, control performance synthesis algorithm for the control system under the consideration of communication, physical cost and online optimization for searching the maximal triggering period are elaborated. A simulation of frequency regulation with a self-triggered control scheme is illustrated in Section 4. Conclusions are given in Section 5.

2. System Model of Power Grid

2.1. Dynamic Model of Power Grid

The electric power network consists of n interconnected power subsystems as shown in Figure 1. It assumes that all power subsystems are same. Each power subsystem consists of a distributed energy source and load, including gas turbine generators, wind power generations and battery arrays [34,35] in the system. These power generating machines supply electric power to meet the demands. Gas turbine, wind power, and battery power output are controllable.

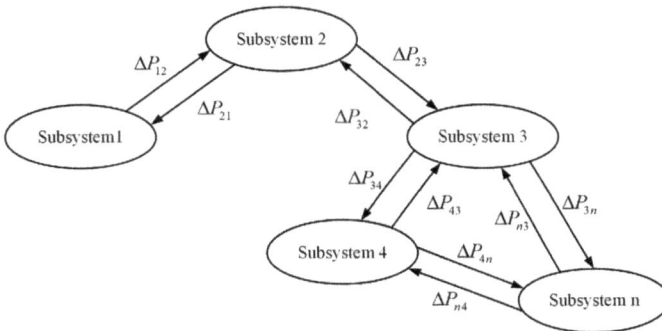

Figure 1. Power grid structure.

Mass loads are considered in the dynamic model of the power grid. Battery electric storage systems and wind power systems, which are connected to the power net by power electronic interface, are controllable. In mathematics, the frequency control method is equivalent to the tie-line bias control (TBC) method as a frequency control in electrical power systems in consideration of tie-line frequency [36–39]. For each subsystem, the block diagram is shown in Figure 2. The meaning of each parameter in the block diagram is given as follows.

- K_i and B_i are TBC gain and frequency bias, respectively.
- T_{gi} and T_{di} are the governor and gas turbine constant, respectively.
- M_i and D are the inertia and damping constant, respectively.
- R_{gi} and T_{ij} are the regulation and synchronizing constant, respectively.
- Δx_{gi} is a governor input of a gas turbine generator.

There are six power notations as follows:

- ΔP_{gi} is an output of the gas turbine generator.
- ΔP_{Wi} is an output of wind power generation.
- ΔP_{Li} is the load fluctuation except controllable load.
- ΔP_{Bi} is an output of the battery electric storage system.
- ΔP_{ji} and ΔP_{ij} is the tie-line power low deviation.
- $\Delta P_{ij} - \Delta P_{ji}$ is the output power of area i, which is delivered to area j.
- ΔP_i in Equation (1) shows the electric power generation of subsystem i and the supply error margin of power consumption.

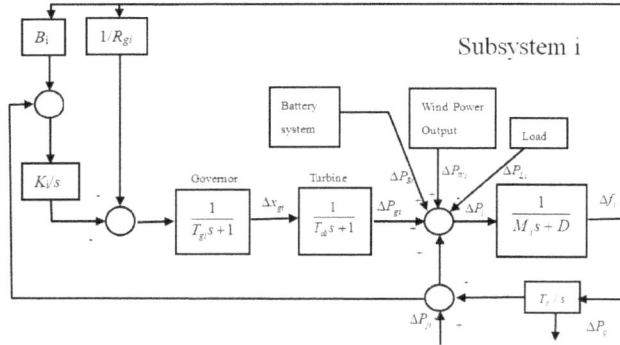

Figure 2. Subsystem structure.

Frequency deviation Δf_i can be calculated by the supply error margin shown in the block diagram. The power ΔP_i for subsystem i is:

$$\Delta P_i = \Delta P_{gi} + \Delta P_{Wi} + \Delta P_{Bi} - \Delta P_{Li} + \Delta P_{ij} - \Delta P_{ji} \tag{1}$$

In a mathematical form, for subsystem i, if the set of neighbored subsystem is denoted as D_i, the dynamics of each subsystem can be described by using continuous time–state equation:

$$\dot{x}_i = A_i x_i + B_i u_i + \sum_{j \in D_i} A_{ji} x_j + E_i w_i \tag{2}$$

$$A_i = \begin{pmatrix} 0 & -\sum_{j \in D_i} T_{ji} & 0 & 0 & 0 \\ 1/M_i & -D_i/M_i & 1/M_i & 0 & 0 \\ 0 & 0 & -1/T_{di} & 1/T_{di} & 0 \\ 0 & -1/(T_{gi}R_{gi}) & 0 & 1/T_{gi} & K_i/T_{gi} \\ 1 & -B_i & 0 & 0 & 0 \end{pmatrix},$$

$$B_i = \begin{pmatrix} 0 & 0 \\ 1/M_i & 1/M_i \\ 0 & 0 \\ 0 & 0 \\ 0 & 0 \end{pmatrix},$$

$$A_{ji} = \begin{pmatrix} 0 & 0 & 0 & 0 & 0 \\ \dfrac{T_{ij}}{M_i \sum_{h \in D_i} T_{ih}} & 0 & 0 & 0 & 0 \\ 0 & 0 & 0 & 0 & 0 \\ 0 & 0 & 0 & 0 & 0 \\ 0 & 0 & 0 & 0 & 0 \end{pmatrix}, j \in D_i.$$

4

where $x = (\Delta P_{outi}, \Delta f_i, \Delta P_{gi}, \Delta x_{gi} U_{AR_i})^T$, and ΔP_{outi}, Δf_i, Δ_{gi}, Δx_{gi}, and U_{AR_i} are output power deviation, frequency deviation, gas generator output power deviation, governor input of gas turbine generator, and regional demand for subsystem i, respectively. w_i is disturbance of the dynamic system, which is bounded. The power output of subsystem i is $\Delta P_{outi} = \sum_{j \in D_i} \Delta P_{ij}$. The tie-line power flow deviation of i is expressed as $\Delta \sum_{j \in D_i} P_{ji} = \sum_{j \in D_i} T_{ij}(\Delta f_j B_i \Delta f_i)$, when the adjoining area is j. The regional demand is defined by $U_{AR_i} = \int AR_i dt$, where $AR_i = \Delta \sum_{j \in D_i} P_{ji}$.

For subsystem i, we assume the L-2 norm of w_i is bounded, and $||w_i||^2 \leq \eta$. The control input u_i for subsystem i is:

$$u_i = -K_i x_i - \sum_{j \in D_i} L_{ji} x_j, \tag{3}$$

where K is the local state feedback gain for control law of Equation (3), $-K_i x_i$ is the local feedback component and $-\sum_{j \in D_i} L_{ji} x_j$ is the control compensation for the neighbors.

2.2. The Self-Triggered Controller

Compared with the distributed control scheme, the advantage of the centralized control scheme in STC is that it reduces the conservativeness of control system, and further decreases the communication cost [40,41]. In order to present the self-triggered control scheme, the power system is formulated as a linear dynamic system in a form of:

$$\dot{x} = Ax + Bu + EW \tag{4}$$

where $x = (x_i)_{n \times 1}, 1 \leq i \leq n$ and $i \in N^+$; $A = (A_{ij})_{n \times n}, A_{ii} = A_i, 1 \leq i, j \leq n$ and $i, j \in N^+$, if no connection exits between subsystem i and j, $A_{ij} = 0$; $B = diag(B_i)_{n \times 1}$ and $E = diag(E_i)_{n \times 1}$, $1 \leq i \leq n$ and $i, j \in N^+$; the control input $u = (u_i)_{n \times 1} = (-L_{ij})_{n \times n}$, $x = -Kx$, $L_{ii} = K_i$, $1 \leq i, j \leq n$ and $i, j \in N^+$, and $L_{ij} = 0$, if there is no connection on subsystems i and j; the disturbance is $W = (w_i)_{n \times 1}, 1 \leq i, j \leq n$ and $i, j \in N^+$. The control objective is to drive state x to origin zero by the linear controller.

In a self-triggered control scheme [42], the local state $x_i(t)$ can be acquired by observers (sensors). However, the information remains within the ith subsystems and is not shared within the system controller unless a pre-calculated triggering time is up, a self-triggered state is reached and a message is sent via data communication links. Thus, the control signal from controller remains constant and may change only after a self-triggered message is received. Self-trigged control promises the reduction of communication cost without sacrificing control performance. For a self-event triggered controller, the dynamic of power system in kth triggering is:

$$\dot{x}(t) = Ax(t) + Bu(t_k) + EW(t), t_k \leq t < t_{k+1} \tag{5}$$

and the control output $u(t_k)$ is:

$$u(t_k) = (-L_{ij})_{n \times n} x(t_k), 1 < i, j < n \quad and \quad i, j \in N^+ \tag{6}$$

As shown in Figure 3, for a self-triggered control scheme, the controller obtains system state x_1, x_2, x_3 from sensors or observers of each subsystem, when the time for triggering t_k is up. Then, the next triggering time t_{k+1} is calculated by using the previous system state $x(t_k)$. Meanwhile, the controller calculates the new control output $u(t_k)$ with the obtained new system state. The new control output $u_1(t) = u_1(t_k), u_2(t) = u_2(t_k), u_3(t) = u_3(t_k)$, where $t_k \leq t < t_{k+1}$ is then applied to its corresponding subsystems. Above all, the procedure of self-triggered control is:

1. First, obtain the system state of each subsystem, when the time for triggering is up;
2. Second, calculate the time for the next triggering;

3. Finally, apply the new control output, which is calculated by using the system state obtained in step 1.

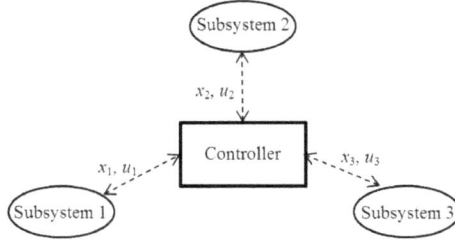

Figure 3. Self-triggered control scheme.

2.3. Exponential Stability and Cost Function

Before elaborating the control scheme for power grid, the exponential stability and cost function are introduced. Exponential stability [43,44] is a kind of asymptotic stability. According to the exponential stability, the state converges to zero with an exponential rate, and $x(t) = x(0)e^{-t}$. If a Lyapunov function satisfies $\kappa_1|x| \leq V(x) \leq \kappa_2|x|$, we have *Proposition 1*.

Proposition 1 [45]. *Let $V : R^n \rightarrow R^+$ be a quadratic Lyapunov candidate function satisfying $V(x) = x^T P x, \forall x \in R^n$, with $P = P^T > 0$. If the condition:*

$$\dot{V}(x) + 2\beta V(x) \leq 0 \tag{7}$$

is satisfied for all trajectories of (4), for a given scalar $\beta > 0$, the system origin is globally β-stable (i.e., there exists a scalar β and α, such that the trajectories satisfy $\|x(t)\| \leq \alpha e^{-\beta t}\|x_0\|$ for any initial condition x_0). Proposition 1 addresses the relationship between the time derivation of Lyapunov function $V(x)$ and the $V(x)$ under the restriction of exponential stability.

As both physical cost and cyber cost are considered, the cost of system consists of two parts. The first part, the physical state cost expressed by the following equation, is a general form that is widely used in optimal control theory [46].

$$p(t) = x(t)^T Q x(t) = \|x(t)\|_Q^2$$

where Q is the weight matrix for state cost. The communication cost as the second part is essentially related to the triggering rate. High triggering rate means high communication bandwidth cost in data transmission within a time unit. Therefore, the communication cost is proportional to the sampling rate.

Definition 1. *At the kth triggering period, the communication cost is defined as:*

$$c(t) = \frac{\varrho}{\tau(k)}.$$

where ϱ is the weight of communication cost, and $\tau(k)$ is the sampling interval between number k sampling and the number $k+1$ sampling.

Integrating physical state cost with communication cost, the cost function is shown as:

$$J = \int_0^{T_f} [q(t) + c(t)]dt = \int_0^{T_f} [\|x(t)\|_Q^2 + \frac{\varrho}{\tau(k)}]dt \tag{8}$$

where T_f is the terminal time. When $t = T_f$, the system state converges to a small value that is close to zero. Moreover, when $t = T_f$, the Lyapunov function value is VT, which is much smaller than the initial Lyapunov function value V_0. c is the number of control actions, which have $T_f = \sum_{i=0}^{n} \tau(i)$. This model considers two parts. The first is a function $\sigma(x(t_k)) = t_{k+1} - t_k$ for calculating self-triggered time, which is adaptive to state $x(t_k)$, to reduce costs in communication; the second is the determination of control gain $K = (-L_{ij})_{n \times n}$. Hence, the cost of system J is guaranteed under a specified upper bound. The following assumptions are made to calculate the upper bound.

Assumption 1. *For a given Lyapunov function $V(x) = x^T P x$, ε and initial state x_0, the system is exponentially stable. The terminal time of control process is T_f defined by $dV(x(T_f))/dt = \nu$, where ν is a small number. The Lyapunov function is $V(x(T_f)) = V_T$, and its initial value is $V(x(0)) = V_0$, which has $V_T < V_0$.*

Assumption 1 is applied to make the value of Lyapunov function close to zero at terminal time T_f, so that the terminal time T_f for cost function can be determined. It should be noted that the variation of the Lyapunov function converges to zero, when the system converges to a stable state. Therefore, ν should be a small number.

3. The Self-Triggered Controller Design

3.1. Function σ for Self Triggering

The communication cost of self-triggered control depends on the triggering rate, which is directly related to the triggering time in self-triggered control. The system calculates the next triggering time and updating control output when the triggering time is up. The function σ for self-triggering is a crucial function. For obtaining the function σ, some important results about exponential stability and input-to-state stability under self-triggered control are introduced. To investigate the relationship between the maximal triggering interval function σ and system performance, exponential stability and input-to-state stability from disturbance W to system state x, Theorem 1 is proposed.

Theorem 1. *For the dynamic system (5), given scalars $\beta > 0, \gamma > 0$, if there exists an $n \times n$ matrix $P = P^T$, a positive scalar γ and a bounded function $\sigma(x) : R^n \rightarrow R_+$ are for all $x \in R^{4n}$ and $\tau \in [0, \sigma(x)]$:*

$$x^T \Phi_{P,\beta}(\tau) x \leq \psi(\tau) \tag{9}$$

where ψ is,

$$\psi = \begin{cases} 0 & \tau \in R_1 \cap R_2 \\ (\gamma/2 - (2\beta + 1)(e^{\alpha\tau} - 1)r\epsilon)n\eta & \tau \in R_1/R_2 \\ (\gamma/2 - r\epsilon e^{\alpha\tau})n\eta & \tau \in R_2/R_1 \\ (\gamma - r\epsilon e^{\alpha\tau}(2\beta + 1)(e^{\alpha\tau} - 1)r\epsilon)n\eta & \tau \in R^+/R_1/R_2 \end{cases}$$

with $R_1 = \{\tau | \gamma/2 - r\epsilon e^{\alpha\tau}\} > 0$, $R_2 = \{\tau | \gamma/2 - (2\beta + 1)(e^{\alpha\tau} - 1)r > 0\}$, $\alpha = \lambda_{max}(A^T + A)$, $\epsilon = \lambda_{max}(P)$, $r = \lambda_{max}(E^T E)$, where,

$$\Phi_{P,\beta}(\tau) = \begin{pmatrix} \Lambda(\tau) \\ I \end{pmatrix}^T \begin{pmatrix} A^T P + PA + 2\beta P & -PBK \\ -K^T B^T P & 0 \end{pmatrix} \begin{pmatrix} \Lambda(\tau) \\ I \end{pmatrix} \tag{10}$$

and,

$$\Lambda(\tau) = I + \int_0^\tau e^{sA} ds(A - BK), \tag{11}$$

then the origin system (2) is global β-stable and input-to-state stability from W to x for any triggering interval $\sigma(x) : R_+ \times R^n \rightarrow R_+$ defines the triggering interval sequence by the law $t_{k+1} = t_k + \sigma(x(t_k))$, $k \in N$. $||x||^2 \leq n(\gamma + \vartheta)\eta/(\epsilon\beta)$ under a given zero initial state $x(t_k) = 0$.

Proof. See Appendix A. $\square$

The key of Theorem 1 is that the function σ must satisfy the inequality (9). However directly solving inequality (9) for obtaining σ is difficult. Approaches, such as in [30,47], are employed to cut the triggering interval (sampling interval) and state space into several sections. Numerical methods, such as the linear matrix inequality (LMI) toolbox, offer a possibility to solve σ. The precision of maximal triggering time depends on the number of sections. Higher precision of maximal triggering time requires more sections. In addition, if the state space dimension is high, the computation cost in the design process dramatically increases. This is not practical in power system control applications with a high state space dimension. Therefore, another algorithm is proposed. The $\sigma(x)$ can be directly computed with a given x on-line by the proposed algorithm. Based on the result of Theorem 1, the following theorem is derived for calculating σ.

Theorem 2. *For the given parameters in Theorem 1, the dynamic system in Equation (5) has a minimal triggering interval for the global state space $\tau_{min} = \min_{x \in R^{4n}} \sigma(x)$, and a maximal sampling interval $\tau_{max} = \max_{x \in R^{4n}} \sigma(x)$. $\forall \tau \in [0, \sigma(x)], x^T \Phi_{P,\beta}(\tau)x \leq 0$, if the $\sigma(x)$ is:*

$$\sigma(x) = \arg_\tau \min_{\tau \in [\tau_{min}, \tau_{max}]} \tau, \tag{12}$$

where τ_{min} and τ_{max} are the lower bound and upper bound of self-triggering interval or sampling interval, respectively. Under the constraint of:

$$x^T \Phi_{P,\beta}(\tau)x = \psi(\tau) \tag{13}$$

then the origin system in Equation (2) is global β-stable with input-to-state stability from W to x.

Proof. From Theorem 1, it is known that the left-hand-side and the right-hand side of inequalities (9) are continuous, and $\sigma(x) \in [\tau_{min}, \tau_{max}]$, thus the maximal triggering interval $\sigma(x)$ must be the minimal root of Equation (13).

From Theorem 2, for a given triggering interval $\tau = t_{k+1} - t_k$, the root of Equation (13) for triggering interval $\sigma(x)$ should be the one which is closest to τ_{min}. Equation (13) can be written as a linear combination of $e^{\lambda_i \tau}$ and $e^{(\lambda_i + \lambda_j)\tau}$, where λ_i and λ_j are the eigenvalues of A. If we denote e^τ by z,

$$\Phi_{P,\beta}(\tau) = \phi_1 + \phi_2 z^{v_1} + \phi_3 z^{v_2} + \ldots + \phi_r z^{v_{r-1}} \tag{14}$$

where ϕ_k is the coefficient matrix of z^{v_k}, v_k is λ_i or $\lambda_i + \lambda_j$, $i \leq n, j \leq n$ and $v_i \neq v_j$. However, the computation cost of directly solving the equation $x^T \Phi_{P,\beta}(\tau)x = 0$ is very high, and it is not practical in online processing for power system control. Instead, the approximate root of this equation can be obtained by the two-point Taylor expansion method. The $\Phi_{P,\beta}(\tau)$ can be expanded into two-point Taylor series in m orders at $z_1 = e^{\tau_{min}}$ and $z_2 = e^{\tau_{max}}$. The approximation of $\Phi_{P,\beta}(\tau)$ is:

$$H(z) = \sum_{k=0}^m \{[a_k(z_1, z_2)(z - z_1) + a_k(z_2, z_1)(z - z_2)](z - z_1)^k (z - z_2)^k\}, \tag{15}$$

where $z = e^\tau$, and $a_n(z_1, z_2)$ is:

$$a_n(z_1, z_2) = \sum_{k=0}^m \{ \frac{(n+k-1)!}{k!(n-k)} \frac{(-1)^{n+1} n \phi^{(n-k)}(z_2) + (-1)^k k \phi^{(n-k)}(z_1)}{n!(z_1 - z_2)^{n+k+1}} \}. \tag{16}$$

and the approximation of the right hand side of Equation (13), denoted $h(z)$, can be obtained similarly as $H(z)$. Above all, for a given state x, we have the following theorem to calculate the $\sigma(x)$. □

Theorem 3. *For the given parameters in Theorem 1, the dynamic system (5), the maximal and minimal triggering intervals* τ_{min}, τ_{max}, *and the maximal approximation error of* $\Phi_{P,\beta}(\tau)$:

$$\varepsilon_1 = \max_{\tau \in [\tau_{min}, \tau_{max}]} eig[\Phi_{P,\beta}(\tau) - H(z)]$$

The maximal approximation error of $h(z)$ is $\varepsilon_2 = \max_{\tau \in [\tau_{min}, \tau_{max}]} \psi_1(\tau) - h_1(z)$, under $\gamma - ree^{\alpha\tau} > 0$, with the approximation expression (15). Then, under the constraint of $\sigma(x) \in [\tau_{min}, \tau_{max}]$, the maximal triggering interval function is chosen by:

$$\sigma(x) = \ln z_c \tag{17}$$

where,

$$z_c = \arg_z \min_{x^T[H(z)+\varepsilon_1 I]x=h_1(z)-\varepsilon_2} |z - z_1| \tag{18}$$

Then, the origin system in Equation (2) is global β-stable with input-to-state stability from W to x.

Proof. See Appendix B. □

For a given state x, z_c can be easily obtained by solving the polynomial in Equation (18). Thus, under the constraint of $\sigma(x) \in [\tau_{min}, \tau_{max}]$, the maximal triggering interval can be obtained by Theorem 3. If the approximation error ε_1 and ε_2 are very small, the approximation maximal triggering interval obtained by applying Theorem 3 should be very close to the actual maximal triggering interval. It should be noted that τ_{min} can be calculated by conventional discrete control theory. However, the maximal sampling interval $\tau_{max} = max_{x \in R^{4n}} \sigma(x)$ cannot be obtained without knowing σ. Thus, τ_{max} is set to be much larger than τ_{min}, $\tau_{max} > \tau_{min}$ in the algorithm to guarantee $\sigma(x) \leq \tau_{max}$.

3.2. The Selection of Feedback Gain for Controller

The system performance to some extent depends on the feedback gain. If the selected feedback gain is not proper for the control system, the τ_{min} in Theorem 1 may not exist, so that β-exponential stability and input-to-state stability cannot be satisfied even under the continuous control. Therefore, the feedback gain should be selected for satisfying the exponential stability and input-to-state stability in continuous control ($\sigma(x) = 0$) first. Otherwise, the cost function should be considered when we select the feedback gain. In dealing with nonlinear control problems, many optimal control theories are proposed, such as Hamilton–Jacobi–Bellman equations, Euler–Lagrange equations and Sontag's formula [44,48]. However, it is difficult to apply those methods to solve this optimal control problem with the cost function described in Section 2. The reason is that solving the Hamilton–Jacobi–Bellman equations and Euler–Lagrange equations is extremely difficult. In addition, Sontag's formula requires a fixed standard form of cost function, which does not coincide with our cost function. It is difficult to obtain the value of cost function directly. Therefore, instead of directly calculating the value of the cost function, inequalities for estimation are derived to select a proper feedback gain for guaranteeing the cost function within an upper bound.

Theorem 4. *For given feedback gain K, γ, β, and $\tau_{min} = min_{x \in R^N} \sigma(x)$ for global state space, which satisfies the condition of Proposition 1 and Assumption 1. Then the cost function (8) can be estimated by:*

$$J < \frac{\lambda_{max}(Q)}{\beta^2 \lambda_{min}(P)} \{\gamma n \eta T'_f \beta - [\gamma n \eta + V_0 \beta](1 - e^{-\beta T'_f})\} + \frac{\varrho T'_f}{\tau_{min}}, \tag{19}$$

where,

$$T'_f = \frac{1}{\beta} ln[(V_0 \beta - \gamma n \eta)/\nu] \geq T_f \tag{20}$$

Proof. From the proof of Theorem 1, the value of the Lyapunov function can be estimated by $V(t) \leq e^{-\beta t}V_0 + \gamma n\eta / \beta (1 - e^{-\beta t})$.

To integrate both sides of the above inequality during 0 and T_f, the cost of state can be estimated by: $\int_0^{T_f} x^T P x \, dt \leq \frac{1}{\beta^2} \{ \gamma n\eta T_f \beta - (\gamma n\eta + V_0\beta)(1 - e^{-\beta T_f}) \}$.

The estimated terminal time T_f' for T_f can be obtained by solving the following inequality: $V(T_f) \leq e^{-\beta T_f}V_0 + \gamma n\eta / \beta (1 - e^{-\beta T_f}) = V_T$, and the time derivatives of V have the inequality $dV(T_f)/dt \leq (\gamma n\eta - V_0\beta)e^{-\beta T_f}$.

Therefore, the result is: $T_f \leq \frac{1}{\beta} ln[(V_0\beta - \gamma n\eta)/v] = T_f'$.

The communication cost can be estimated by $c \leq \varrho T_f / \tau_{min} \leq \varrho T_f' / \tau_{min}$. Above all, the value of the cost function at terminal time T_f can be estimated by summing the estimation of state cost and communication cost. □

For a given P for Lyapunov function, the decay rate β, and γ for the inhibition of disturbance effect, the minimal sampling interval τ_{min} and the feedback gain K can be figured out by conventional robust discrete control technique. Then, the upper bound of cost function J can be estimated by a given Lyapunov function value V_0 and V_T in initial time and terminal time by Theorem 4. The objective is to minimize the upper bound of J, so that the value of the cost function J can be guaranteed on a required level. Therefore, the feedback gain K, decay rate β, parameter γ for input-to-state stability, minimal sampling interval τ_{min}, and Lyapunov function parameter P, can be estimated by minimizing the upper bound of J. If we denote the right-hand side of inequality (19) by χ, it is:

$$\{K, \beta, P, \gamma\} = \arg \min_{\{K,\beta,P,\gamma\}} \chi(K, \beta, P, \gamma) \tag{21}$$

Some numerical optimization methods can be applied to solve this optimization problem. Let $\theta = \{K, \beta, P, \gamma\}$. The optimization process in one iteration can be depicted as follows:

1. For a given θ, τ_{min} is calculated by conventional discrete robust control technique;
2. The upper bound of cost function χ is obtained by Theorem 4, and R_k can be calculated;
3. Update θ by numerical optimization algorithms (such as GA optimization algorithm) with χ obtained in previous step;
4. Return to the first step until the stop criteria is satisfied.

The iteration stop condition depends on the value of χ and the iteration count and satisfies the design requirements. With the smallest upper bound of cost, the performance of control is guaranteed. The cost consists of state cost and communication cost. Thus, the Pareto Frontier curve may be calculated for generality. Then the optimal parameter θ can be easily obtained for any ϱ.

3.3. Event-Triggered Control Algorithm

The self-triggered control is divided into two phrases. The first phrase is about parameter design, and the second phrase is online computing of triggering interval $\sigma(x(t_k))$. First, for a given power grid dynamic model (4) and Q, the τ_{min} is calculated. The best value of θ is designed by Theorem 4. Optimal searching algorithms consider both state and communication cost in cost function. If the feedback with gain K does not have a solution with given γ and β, β and γ should decrease and increase respectively until there is a solution of K. After θ, τ_{min} and τ_{max} are selected, the approximation error ε_1, ε_2 and ε_3 are calculated. Meanwhile, the two-point Taylor expansion of Φ, ψ_1 and ψ_2 can be obtained by software such as Mathematica.

Second, the triggering interval is calculated online. The power system obtains new system state x_k from sensors, when a self-triggered time is up. Then, the time $t_{k+1} = t_k + \sigma(x(t_k))$ is calculated for the next triggering. For a system state x_k, the coefficient of polynomial respect to z is derived by $x^T H(z)x$. Then z_c is solved using Equation (18) by inverse iteration. Therefore, the triggering interval can be obtained by z_c and (17). The new control output $u = -Kx_{(t_k)}$ is calculated and applied to actuators. At last, the control output is updated at the next triggering time t_{k+1}.

4. Simulation Results

In this section, a simulation of power frequency control with distributed energy source demonstrates the effectiveness and advantages of our proposed control method. The subsystem frequencies are controlled by our proposed controller, which can save more costs under the consideration of communication and system state. Meanwhile, a comparison is carried out to verify the benefit on the control of power system. The simulation is performed in MATLAB 2010b.

We consider the electrical power network shown in Figure 4. Three subsystems are in the power system. It assumes that the composition of three electric power subsystems is same, which is illustrated in Figure 2. There are gas turbine generators, wind power generations and battery arrays in the system. Power supply is done to the electric power demand with these power generating machines. The gas turbine, wind power and battery power output are controllable.

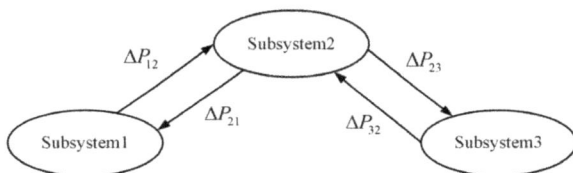

Figure 4. Power system structure for simulation

The parameters of each subsystem are given in Table 1.

Table 1. Parameter set.

Parameters ($)	Symbols (Unit) ($)	Values		
		Subsystem 1	Subsystem 2	Subsystem 3
Inertia constant	M (puMw s/Hz)	0.20	0.14	0.16
Damping constant	D (puMw/Hz)	0.26	0.26	0.23
Governor constant	T_g (s)	0.20	0.20	0.12
Gas turbine constant	T_d (s)	5.0	4.5	5.0
Regulation constant	R_g (Hz/pu Mw)	2.5	2.5	1.5
Synchronizing constant	T_{ij} (pu Mw)	0.50	0.5	0.5
TBC gain	K_i	0.1	0.08	0.1
Frequency bias	B_i (Mw/Hz)	0.1	0.1	0.08

The maximal norm of disturbance assumes to be 0.1, and the weight ϱ for communication cost is set to 0.1. The contour map of cost upper bound with respect to exponential stability parameter β and input-to-state stability parameter γ is illustrated in Figure 5, and the relationship between β and τ_{min} is shown in Figure 6.

According to the cost upper bound contour map, if the parameter β is selected to be very low (lower than 0.02), then the effect of parameter γ is not significant. Meanwhile, it shows that the faster convergence rate (larger β) makes the physical cost (state cost) lower. However, according to the relationship between the τ_{min} and β illustrated in Figure 6, large β may bring a higher communication cost, because the controller has to reduce the triggering interval to satisfy higher physical cost requirements (faster rate of convergence). Thus, a tradeoff should be made to reduce the total cost. After solving the optimization problem described in (21) with MATLAB Optimization ToolBox, β and γ are obtained as 0.12 and 0.11, respectively. In addition, the feedback gain K and Lyapunov function parameter P are also calculated by the robust control design algorithm and LMI toolbox. The initial state is set to 1, and the terminal time is calculated to be 10. The simulation is divided into two groups. The first group is the proposed algorithm with approximation method to calculate the triggering interval. The second method is the method proposed in [42], which is a widely used method in self-triggered control. We call it the "conventional method" in the following. The simulation results and comparisons are illustrated in Figures 7–11.

According to the simulation results in Figures 7–11, we know that the control method can make all states converge to zero with exponential rate β. Moreover, the curve with proposed control method converges slightly faster than the conventional method in the beginning, especially in the curve of Δf and Δx_g.

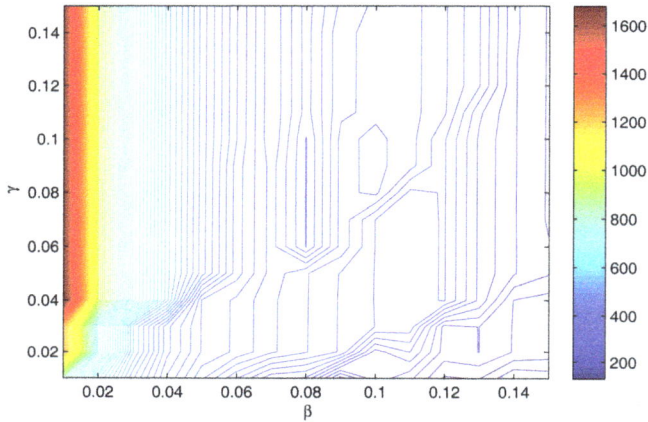

Figure 5. Contour map of cost upper bound with respect to β and γ.

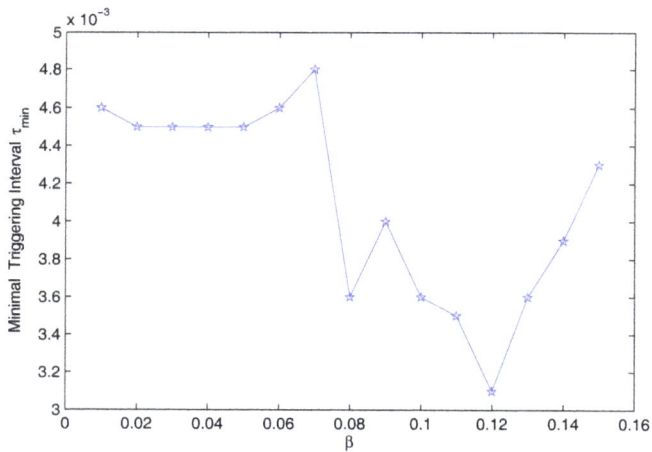

Figure 6. The minimal triggering interval curve with respect to β.

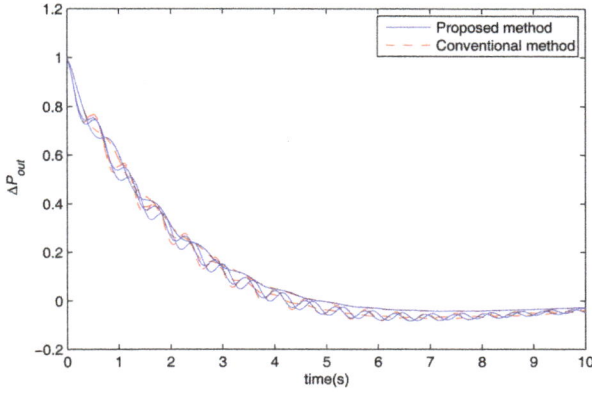

Figure 7. The output power deviation of each subsystem P_{out}.

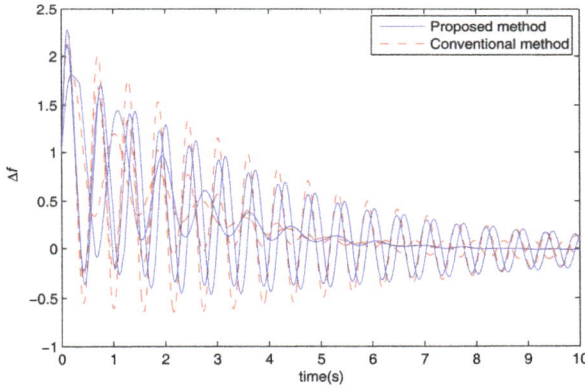

Figure 8. The frequency deviation Δf.

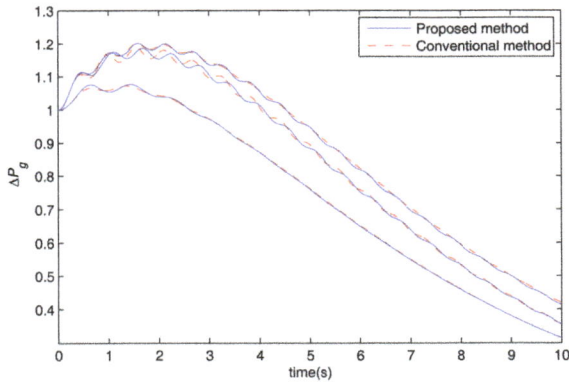

Figure 9. The power output deviation of the gas turbine generator P_g.

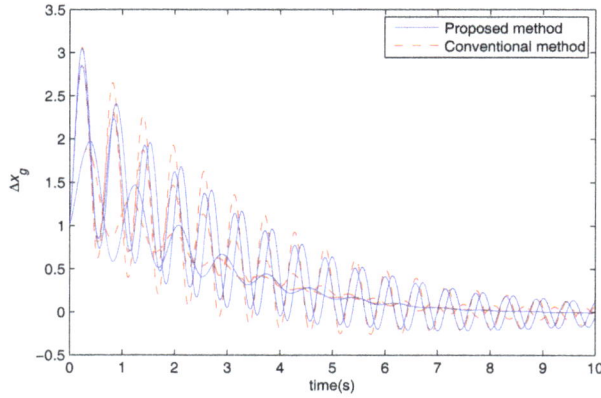

Figure 10. The governor input of the gas turbine generator Δx_g.

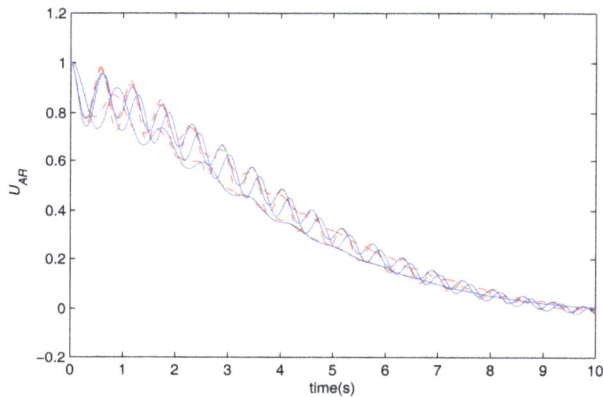

Figure 11. The regional demand U_{AR}.

At last, they are in the same magnitude. Above all, under the same exponential convergence rate β and input-to-state stability parameter γ requirement, the convergence rate or the physical cost seems almost same and satisfies the performance. The proposed method is effective in the frequency control application of power system. However, the communication should be considered in our proposed control algorithm. Therefore the comparison of communication cost is illustrated. As the sampling rate of classical control method is fixed, it causes the communication cost of the classical control method to be much higher than for the STC method [49,50]. We only compare the proposed STC method with the conventional STC method in Figure 12. The total cost is illustrated in Figure 13.

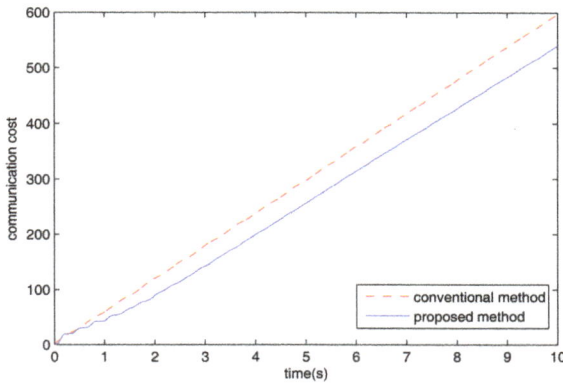

Figure 12. Communication cost comparison.

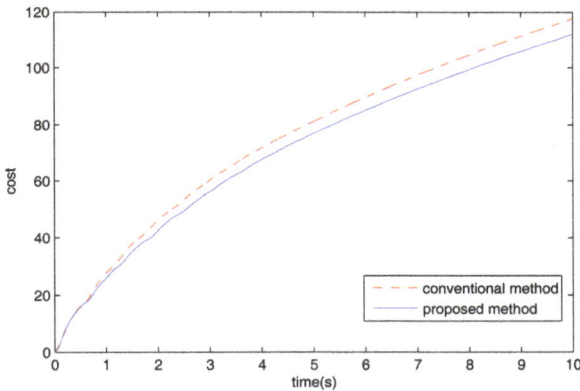

Figure 13. Cost curve comparison.

According to the communication cost comparison in Figure 12, it is known that the communication cost of the proposed self-triggered control method is lower than that of the conventional self-triggered control method. At the end of the simulation time, it costs about 530 control actions using the proposed self-triggered control method. In comparison, it costs about 600 control actions using conventional self-triggered control method. According to the definition of communication cost, the system takes 530 control actions for sensor data acquisition to compute control output with the proposed method, and needs 600 control actions using the conventional method. Based on the total cost curve comparison, the proposed self-triggered method costs less than the conventional method. At the end of simulation, the conventional method costs about 120 control actions, and the proposed method costs about 110 control actions. The cost of the proposed method is not much lower than for the conventional method. When the communication cost is high, bandwidth is limited, or communication network is publicly shared, the triggering time of proposed method further improves. Besides, the computation cost of calculating triggering time in the proposed method is reduced by the approximation method. Above all, the proposed method is better than the conventional method in frequency control application of power system, at least under this situation.

5. Conclusions

In this paper, a novel self-triggered control method is proposed and applied to the frequency control application of the power system. The power system dynamic model consists of multiple subsystems that have distributed energy sources. Physical cost and communication cost as two parameters of cost function are considered in the proposed model. On one hand, the equation for solving the triggering time is derived by the definition of exponential stability and the input of state stability, and an approximation algorithm is proposed to reduce computation costs. On the other hand, the upper bound of cost is derived. The feedback gain and parameters are selected, according to optimizing the upper bound of the cost. Thus, the system cost can be guaranteed under a required level. At last, a simulation of power system frequency control is carried out to demonstrate that the proposed method is effective and can save more costs than the conventional method. Compared with the distributed control scheme, the advantage of the centralized control scheme in STC is that it reduces the conservativeness of the control system, and further decreases the communication cost. Meanwhile, it may require more computational resources and time. Additionally, more communication networks are needed in the centralized control scheme. In future research, the application of self-triggered control method in the transient control of power system for avoiding cascade failure will be investigated.

Acknowledgments: This work was supported in part by the National Natural Science Foundation of China under Grant 61633005. This research is also funded by Chongqing Natural Science Foundation Grant cstc2016jcyjA0428 and Grant cstc2011jjA40013. This project was granted Fundamental Research Funds for the Central Universities cstc2016jcyjA0428.

Author Contributions: Zhiqin Zhu and Jian Sun conceived and designed the experiments; Zhiqin Zhu and Jian Sun performed the experiments; Jian Sun and Guanqiu Qi analyzed the data; Zhiqin Zhu contributed reagents/materials/analysis tools; Zhiqin Zhu and Jian Sun wrote the paper; Yi Chai and Yinong Chen provided technical support and revised the paper.

Conflicts of Interest: The authors declare no conflict of interest.

Appendix A. Proof of Theorem 1

For $t \in [t_k, t_{k+1}]$ and $t = t_k + \tau$, it is known that $x(t) = \zeta(\tau)x(t_k) + \xi(\tau)$, where,

$$\zeta(\tau) = I + \int_0^\tau e^{As}ds(A - BK), \xi(\tau) = \int_0^\tau e^{As}EW(s)ds$$

Therefore for the Lyapunov function

$$V(x(t)) = x(t)^T Px(t) = [\zeta(\tau)x(t_k) + \xi(\tau)]^T P[\zeta(\tau)x(t_k) + \xi(\tau)] \leq x(x_k)^T \zeta(\tau)^T P\zeta(\tau)x(x_k) + \xi^T P\xi.$$

We denote:

$$V'(x(t_k), \tau) = x(x_k)^T \zeta(\tau)^T P\zeta(\tau)x(x_k) + \xi^T P\xi.$$

We define ϖ as the mean of EW(s) in the paper. For the given $\beta > 0$ and $\gamma > 0$, if,

$$\dot{V'} \leq -2\beta \dot{V'} + \gamma/2||W||^2 + \gamma/2||\varpi||^2, \tag{A1}$$

where $\xi = \int_0^\tau e^{As}ds\varpi$, then,

$$V'(x(t_k), \tau) \leq e^{-\beta\tau}V'(x(t_k), 0) + \int_0^\tau (\gamma/2e^{-\beta s}||W(s)||^2 + \gamma/2||\varpi||^2)ds,$$

so that,

$$V(x(t)) \leq V'(x(t_k), \tau) \leq e^{-\beta\tau} V(x(t_k)) + \int_0^\tau \gamma/2 e^{-\beta s} ||W(s)||^2 ds + \int_0^\tau \gamma/2 e^{-\beta s} ||\omega(s)||^2 ds.$$

which means the dynamic system is of β-exponential stability. When $||\omega||^2 \leq n\eta$, we have $||x||^2 \leq n\gamma\eta/(\epsilon\beta)$ with initial state $x(t_k) = 0$. The inequality (A1) is:

$$d[x(t_k)^T\zeta(\tau)^T P\zeta(\tau)x(t_k)]/dt + 2\beta x(t_k)^T\zeta(\tau)^T P\zeta(\tau)x(t_k) \leq -d(\xi^T P\xi)/dt - 2\beta(\xi^T P\xi) + \gamma/2||W||^2 + \gamma/2||\omega||^2, \quad \text{(A2)}$$

and the inequality is:

$$-d(\xi^T P\xi)/dt - 2\beta(\xi^T P\xi) + \gamma/2||W||^2 \geq \dot{\xi}^T P\dot{\xi} - (2\beta+1)(\xi^T P\xi) + \gamma/2||W||^2 \geq$$
$$(\gamma/2 - ree^{\alpha\tau})||W||^2 + [\gamma/2 - (2\beta+1)(e^{\alpha\tau}-1)r]\epsilon||\omega||^2.$$

It should be noted that the left-hand side of inequality (A2) is equivalent to the left-hand side of inequality (9). If:

$$d[x(t_k)^T\zeta(\tau)^T P\zeta(\tau)x(t_k)]/dt + 2\beta x(t_k)^T\zeta(\tau)^T P\zeta(\tau)x(t_k) \leq$$
$$(\gamma/2 - ree^{\alpha\tau})||W||^2 + [\gamma/2 - (2\beta+1)(e^{\alpha\tau}-1)r]\epsilon||\omega||^2,$$

then the inequality (A1) satisfies. When $\gamma/2 - ree^{\alpha\tau} > 0$, and $\gamma/2 - (2\beta+1)(e^{\alpha\tau}-1)r > 0$, the above inequality holds if:

$$d[x(t_k)^T\zeta(\tau)^T P\zeta(\tau)x(t_k)]/dt + 2\beta x(t_k)^T\zeta(\tau)^T P\zeta(\tau)x(t_k) \leq 0.$$

When $\gamma/2 - ree^{\alpha\tau} > 0$, and $\gamma/2 - (2\beta+1)(e^{\alpha\tau}-1)r \leq 0$, the above inequality holds if:

$$d[x(t_k)^T\zeta(\tau)^T P\zeta(\tau)x(t_k)]/dt + 2\beta x(t_k)^T\zeta(\tau)^T P\zeta(\tau)x(t_k) \leq [\gamma/2 - (2\beta+1)(e^{\alpha\tau}-1)r\epsilon]n\eta.$$

When $\gamma/2 - ree^{\alpha\tau} \leq 0$, and $\gamma/2 - (2\beta+1)(e^{\alpha\tau}-1)r > 0$, the above inequality holds if:

$$d[x(t_k)^T\zeta(\tau)^T P\zeta(\tau)x(t_k)]/dt + 2\beta x(t_k)^T\zeta(\tau)^T P\zeta(\tau)x(t_k) \leq (\gamma/2 - ree^{\alpha\tau})n\eta.$$

When $\gamma/2 - ree^{\alpha\tau} \leq 0$, and $\gamma/2 - (2\beta+1)(e^{\alpha\tau}-1)r \leq 0$, the above inequality holds if:

$$d[x(t_k)^T\zeta(\tau)^T P\zeta(\tau)x(t_k)]/dt + 2\beta x(t_k)^T\zeta(\tau)^T P\zeta(\tau)x(t_k) \leq$$
$$(\gamma/2 - ree^{\alpha\tau})n\eta + [\gamma/2 - (2\beta+1)(e^{\alpha\tau}-1)r\epsilon]n\eta.$$

Appendix B. Proof of Theorem 3

Since the approximation errors are given, under $\tau \in [\tau_{min}, \tau_{max}]$, we have $\Phi_{P,\beta}(\tau) - [H(z) + \epsilon I] \leq 0$. Hence,

$$x^T\Phi_{P,\beta}(\tau)x \leq x^T H(z)x + x^T \varepsilon_1 Ix. \quad \text{(A3)}$$

As we know, $|\psi_1(\tau) - h_1(z)| \leq \varepsilon_2$ and $|\psi_2(\tau) - h_2(z)| \leq \varepsilon_3$ where $\tau \in [\tau_{min}, \tau_{max}]$, thus if $x^T\Phi_{P,\beta}(\tau)x \leq x^T H(z)x + x^T \varepsilon_1 Ix \leq h_1(\tau) - \varepsilon_2$ then $x^T\Phi_{P,\beta}(\tau)x \leq \psi_1(\tau) \leq \psi(\tau)$.

Therefore, the z_c calculated from Equation (18), makes the inequality (9) in Theorem 1.

References

1. Zhang, X.M.; Han, Q.L.; Yu, X. Survey on Recent Advances in Networked Control Systems. *IEEE Trans. Ind. Inform.* **2016**, *12*, 1740–1752.

2. Zhang, X.M.; Han, Q.L.; Zhang, B.L. An Overview and Deep Investigation on Sampled-Data-Based Event-Triggered Control and Filtering for Networked Systems. *IEEE Trans. Ind. Inform.* **2017**, *13*, 4–16.

3. Li, H.; Qiu, H.; Yu, Z.; Li, B. Multifocus image fusion via fixed window technique of multiscale images and non-local means filtering. *Signal Process.* **2017**, *138*, 71–85.

4. Li, H.; Liu, X.; Yu, Z.; Zhang, Y. Performance improvement scheme of multifocus image fusion derived by difference images. *Signal Process.* **2016**, *128*, 474–493.

5. Li, H.; Li, X.; Yu, Z.; Mao, C. Multifocus image fusion by combining with mixed-order structure tensors and multiscale neighborhood. *Inf. Sci.* **2016**, *349–350*, 25–49.

6. Sun, J.; Zheng, H.; Demarco, C.L.; Chai, Y. Energy Function-Based Model Predictive Control With UPFCs for Relieving Power System Dynamic Current Violation. *IEEE Trans. Smart Grid* **2016**, *7*, 2933–2942.

7. Anta, A.; Tabuada, P. To Sample or not to Sample: Self-Triggered Control for Nonlinear Systems. *IEEE Trans. Autom. Control* **2010**, *55*, 2030–2042.

8. Anta, A.; Tabuada, P. Exploiting Isochrony in Self-Triggered Control. *IEEE Trans. Autom. Control* **2012**, *57*, 950–962.

9. Miskowicz, M. *Event-Based Control and Signal Processing*; CRC Press: Boca Raton, FL, USA, 2015.

10. Zhu, Z.; Qi, G.; Chai, Y.; Li, P. A Geometric Dictionary Learning Based Approach for Fluorescence Spectroscopy Image Fusion. *Appl. Sci.* **2017**, *7*, 161.

11. Zhu, Z.; Qi, G.; Chai, Y.; Chen, Y. A Novel Multi-Focus Image Fusion Method Based on Stochastic Coordinate Coding and Local Density Peaks Clustering. *Future Internet* **2016**, *8*, 53.

12. Wang, K.; Qi, G.; Zhu, Z.; Chai, Y. A Novel Geometric Dictionary Construction Approach for Sparse Representation Based Image Fusion. *Entropy* **2017**, *19*, 306.

13. Sun, J.; Hu, Y.; Chai, Y.; Ling, R.; Zheng, H.; Wang, G.; Zhu, Z. L-infinity event-triggered networked control under time-varying communication delay with communication cost reduction. *J. Frankl. Inst.* **2015**, *352*, 4776–4800.

14. Sun, J.; Zheng, H.; Chai, Y.; Hu, Y.; Zhang, K.; Zhu, Z. A direct method for power system corrective control to relieve current violation in transient with UPFCs by barrier functions. *Int. J. Electr. Power Energy Syst.* **2016**, *78*, 626–636.

15. Velasco, M.; Fuertes, J.M.; Marti, P. The self triggered task model for real-time control systems. In Proceedings of the 24th IEEE Real-Time Systems Symposium, Cancun, Mexico, 3–5 December 2003; pp. 67–70.

16. Tsai, W.T.; Luo, J.; Qi, G.; Wu, W. Concurrent Test Algebra Execution with Combinatorial Testing. In Proceedings of the 2014 IEEE 8th International Symposium on Service Oriented System Engineering, Oxford, UK, 7–11 April 2014; pp. 35–46.

17. Wu, W.; Tsai, W.T.; Jin, C.; Qi, G.; Luo, J. Test-Algebra Execution in a Cloud Environment. In Proceedings of the 2014 IEEE 8th International Symposium on Service Oriented System Engineering, Oxford, UK, 7–11 April 2014; pp. 59–69.

18. Diaz-Cacho, M.; Delgado, E.; Barreiro, A.; Falcón, P. Basic Send-on-Delta Sampling for Signal Tracking-Error Reduction. *Sensors* **2017**, *17*, 312.

19. Zhu, Z.; Chai, Y.; Yin, H.; Li, Y.; Liu, Z. A novel dictionary learning approach for multi-modality medical image fusion. *Neurocomputing* **2016**, *214*, 471–482.

20. Rafael, S.; Sebastián, D.; Raquel, D.; Ernesto, F. Event-Based Control Strategy for Mobile Robots in Wireless Environments. *Sensors* **2015**, *15*, 30076–30092.

21. Tsai, W.T.; Qi, G. DICB: Dynamic Intelligent Customizable Benign Pricing Strategy for Cloud Computing. In Proceedings of the 2012 IEEE Fifth International Conference on Cloud Computing, Honolulu, HI, USA, 24–29 June 2012; pp. 654–661.

22. Tsai, W.T.; Qi, G.; Chen, Y. A Cost-Effective Intelligent Configuration Model in Cloud Computing. In Proceedings of the 2012 32nd International Conference on Distributed Computing Systems Workshops, Macau, China, 18–21 June 2012; pp. 400–408.

23. Tsai, W.T.; Qi, G.; Chen, Y. Choosing cost-effective configuration in cloud storage. In Proceedings of the 2013 IEEE Eleventh International Symposium on Autonomous Decentralized Systems (ISADS), Mexico City, Mexico, 6–8 March 2013; pp. 1–8.

24. Pandey, S.K.; Mohanty, S.R.; Kishor, N.; Catalão, J.P.S. Frequency regulation in hybrid power systems using particle swarm optimization and linear matrix inequalities based robust. *Int. J. Electr. Power Energy Syst.* **2014**, *63*, 887–900.

25. Bhatt, P.; Ghoshal, S.P.; Roy, R. Coordinated control of TCPS and SMES for frequency regulation of interconnected restructured power systems with dynamic participation from DFIG based wind farm. *Renew. Energy* **2012**, *40*, 40–50.

26. Mohanty, S.R.; Kishor, N.; Ray, P.K. Robust H-infinite loop shaping controller based on hybrid PSO and harmonic search for frequency regulation in hybrid distributed generation system. *Int. J. Electr. Power Energy Syst.* **2014**, *60*, 302–316.

27. Dai, Y.; Xu, Y.; Dong, Z.Y.; Wong, K.P.; Zhuang, L. Real-time prediction of event-driven load shedding for frequency stability enhancement of power systems. *IET Gen. Transm. Distrib.* **2012**, *6*, 914–921.

28. Cao, J.; Du, W.; Wang, H.; Chen, Z.; Li, H.F. A Novel Emergency Damping Control to Suppress Power System Inter-Area Oscillations. *IEEE Trans. Power Syst.* **2013**, *28*, 3165–3173.

29. Xu, Y.; Dong, Z.Y.; Luo, F.; Zhang, R.; Wong, K.P. Parallel-differential evolution approach for optimal event-driven load shedding against voltage collapse in power systems. *IET Gen. Transm. Distrib.* **2014**, *8*, 651–660.

30. Fiter, C.; Hetel, L.; Perruquetti, W.; Richard, J.-P. A state dependent sampling for linear state feedback. *Automatica* **2012**, *48*, 1860–1867.

31. Mazo, M.; Tabuada, P. Decentralized Event-Triggered Control Over Wireless Sensor/Actuator Networks. *IEEE Trans. Autom. Control* **2011**, *56*, 2456–2461.

32. Mazo, M., Jr.; Anta, A.; Tabuada, P. An ISS self-triggered implementation of linear controllers. *Automatica* **2010**, *46*, 1310–1314.

33. Heemels, W.P.M.H.; Donkers, M.C.F.; Teel, A.R. Periodic Event-Triggered Control for Linear Systems. *IEEE Trans. Autom. Control* **2013**, *58*, 847–861.

34. Namerikawa, T.; Kato, T. Distributed load frequency control of electrical power networks via iterative gradient methods. In Proceedings of the 50th IEEE Conference on Decision and Control and European Control(CDC-ECC), Orlando, FL, USA, 12–15 December 2011; pp. 7723–7728.

35. Keerqinhu; Qi, G.; Tsai, W.; Hong, Y.; Wang, W.; Hou, G.; Zhu, Z. Fault-Diagnosis for Reciprocating Compressors Using Big Data. In Proceedings of the Second IEEE International Conference on Big Data Computing Service and Applications (BigDataService), Oxford, UK, 29 March–1 April 2016; pp. 72–81.

36. Kundur, P. *Power System Stability and Control*; McGraw-Hill Education: New York, NY, USA, 1994.

37. Xiao, B.; Lam, H.K.; Li, H. Stabilization of Interval Type-2 Polynomial-Fuzzy-Model-Based Control Systems. *IEEE Trans. Fuzzy Syst.* **2017**, *15*, 205–217.

38. Tsai, W.T.; Qi, G.; Hu, K. Autonomous Decentralized Combinatorial Testing. In Proceedings of the 2015 IEEE Twelfth International Symposium on Autonomous Decentralized Systems, Taichung, Taiwan, 25–27 March 2015; pp. 40–47.

39. Tsai, W.T.; Qi, G. Integrated Adaptive Reasoning Testing Framework with Automated Fault Detection. In Proceedings of the 2015 IEEE Symposium on Service-Oriented System Engineering, San Francisco Bay, CA, USA, 30 March–3 April 2015; pp. 169–178.

40. Tsai, W.T.; Qi, G. Integrated fault detection and test algebra for combinatorial testing in TaaS (Testing-as-a-Service). *Simul. Model. Pract. Theory* **2016**, *68*, 108–124.

41. Zuo, Q.; Xie, M.; Qi, G.; Zhu, H. Tenant-based access control model for multi-tenancy and sub-tenancy architecture in Software-as-a-Service. *Front. Comput. Sci.* **2017**, *11*, 465–484.

42. Gommans, T.; Antunes, D.; Donkers, T.; Tabuada, P.; Heemels, M. Self-triggered linear quadratic control. *Automatica* **2014**, *50*, 1279–1287.

43. Guo, L.; Ljung, L. Exponential stability of general tracking algorithms. *IEEE Trans. Autom. Control* **1995**, *40*, 1376–1387.

44. Xiao, B.; Lam, H.K.; Song, G.; Li, H. Output-feedback tracking control for interval type-2 polynomial fuzzy-model-based control systems. *Neurocomputing* **2017**, *242*, 83–95.

45. Fiter, C.; Hetel, L.; Perruquetti, W.; Richard, J.-P. State dependent sampling: An LMI based mapping approach. In Proceedings of the 18th IFAC World Congress (IFAC'11), Milano, Italy, 28 August–2 September 2011.

46. Acikmese, B.; Carson, J.M.; Blackmore, L. Lossless Convexification of Nonconvex Control Bound and Pointing Constraints of the Soft Landing Optimal Control Problem. *IEEE Trans. Control Syst. Technol.* **2013**, *21*, 2104–2113.

47. Zhu, Z.; Qi, G.; Chai, Y.; Yin, H.; Sun, J. A Novel Visible-infrared Image Fusion Framework for Smart City. *Int. J. Simul. Process Model.* **2017**, in press.

48. Primbs, J.A.; Nevistic, V.; Doyle, J.C. Nonlinear Optimal Control: A Control Lyapunov Function and Receding Horizon Perspective. *Asian J. Control* **1999**, *1*, 14–24.
49. Tsai, W.; Qi, G.; Zhu, Z. Scalable SaaS Indexing Algorithms with Automated Redundancy and Recovery Management. *Int. J. Softw. Inform.* **2013**, *7*, 63–84.
50. Wang, S.; Zhang, P.; Fan, Y. Centralized event-triggered control of multi-agent systems with dynamic triggering mechanisms. In Proceedings of the 2015 27th Chinese Control and Decision Conference, Qingdao, China, 23–25 May 2015; pp. 2183–2187.

applied
sciences

MDPI

Article

Multi-Objective Optimization of Voltage-Stability Based on Congestion Management for Integrating Wind Power into the Electricity Market

Jin-Woo Choi and Mun-Kyeom Kim *

Department of Energy System Engineering, Chung-Ang University, 84 Heukseok-ro, Dongjak-gu, Seoul 156-756, Korea; spjw11@naver.com
* Correspondence: mkim@cau.ac.kr; Tel./Fax: +82-2-5271-5867

Academic Editors: Amjad Anvari-Moghaddam and Josep M. Guerrero
Received: 7 April 2017; Accepted: 30 May 2017; Published: 2 June 2017

Abstract: This paper proposes voltage-stability based on congestion management (CM) for electricity market environments and considers the incorporation of wind farms into systems as well. A probabilistic voltage-stability constrained optimal power flow (P-VSCOPF) is formulated to maximize both social welfare and voltage stability. To reflect the probabilistic influence of CM in the presence of wind farms on voltage stability, Monte Carlo simulations (MCS) are used to analyze both the system load and the wind speed from their probability distribution functions. A multi-objective particle-swarm optimization (MOPSO) algorithm is implemented to solve the P-VSCOPF problem. A contingency analysis based on the voltage stability index (VSI) for line outages is employed to find the vulnerable line of congestion in power systems. The congestion distribution factor (CDF) is also used to find the optimal location of a wind farm in CM. The optimal pricing expression, which is obtained, with respect to preserving voltage stability, by calculating both the locational marginal prices (LMPs) and the nodal congestion prices (NCPs), is demonstrated in terms of congestion solutions. Simultaneously, the voltage stability margin (VSM) is considered within the CM framework. The proposed approach is implemented on a modified IEEE 24-bus system, and the results obtained are compared with the results of other optimal power flow methods.

Keywords: congestion management; probabilistic voltage-stability-constrained optimal power flow; congestion distribution factor; voltage stability margin; multi-objective particle swarm optimization; wind farm

1. Introduction

Congestion management (CM) is one of the most critical transmission problems of open access environments [1]. Congestion occurs in a power market if the transmission system is incapable of adjusting all of the desired transactions because of the existence of system violations [2]. System congestion may also threaten the reliability of the power system by making it more vulnerable to sudden disturbances. It may also impede market efficiency, forcing consumers to reduce their electric power consumptions because of the rises in market prices. As the organizations responsible for maintaining power system security efficiently, independent system operators (ISOs) have to mitigate congestion problems by using market-based tools and/or effective operating facilities [3]. Market-based techniques using both locational marginal prices (LMPs) and nodal congestion prices (NCPs) have been proposed to manage and relieve transmission congestion [4]. The LMP is the generation marginal cost of meeting both the power demands and the transmission losses at a specific node, and the NCP is the cost of satisfying network security parameter limits. Consequently, the market price structure both provides the marginal cost of generating units and indicates the system security cost of the power network. Thus, it is currently used by ISOs as well as many researchers as CM pricing [5,6]. In addition, the influence of CM can be analyzed through the results obtained from LMPs and NCPs.

Recently, wind energy has emerged as a critical candidate for bridging the gap between global power supply and demand [7]. Wind energy is primarily considered as a sustainable method of mitigating severe problems that result from the use of fossil fuels, such as market volatility, social conflict, and global warming [8,9]. Hence, special attention should be paid to wind energy sources in the CM of the electricity market. CM is essential to the efficient operation of any electricity market, with a special emphasis on wind energy. Assume that several buses in a power system each have strong potential for wind installation. An ISO should consider several factors in determining the optimal placements for wind farms in order to alleviate transmission congestion via additional power injection. CM that uses proper power injection reduces both the component of the LMP associated with the congestion price and reduces transmission losses. Several CM problems involving renewable energy sources have been analyzed in the literature [10–12]. In [10], the authors proposed a new optimal model of congestion management with an emphasis on the promotion of renewable energy sources in a competitive electricity market. In [11], the CM problem incorporated a wind farm based on the sensitivity factor. In [12], the authors addressed transmission congestion relief by considering both the sizes and the sites of new renewable energy sources. However, the impact of renewable energy congestion relief on the probabilistic approach was not considered. Wind behavior is often unpredictable, as it is a stochastic phenomenon. In particular, wind speed is highly dependent upon the weather conditions, geographical region, and season. Wind farms, which deploy many wind turbine generators to harness wind energy for electricity production, have a variable power outputs due to variations in wind speed [13]. Therefore, the rapid global growth of wind power capacity may increase the uncertainties for congestion problems. A reliable probabilistic method that can consider random wind speeds must be found to solve congestion problems.

CM is essentially an optimization problem with an exponential number of constraints that can be generally described as an optimal power flow (OPF) problem whose objective function is the maximization of social welfare and whose constraints are the load flow equations and the operation limitations [14]. In [15], the authors applied CM to the adjustments of power transfers in transmission lines based on a transmission congestion penalty-factor. A CM method based on OPF, presented in [16], relieved congested transmission lines by using both power generation rescheduling and load curtailment. In [17], the authors analyzed both the enhancement of voltage loadability and the transmission mechanisms of line outage contingencies in a smart power network and found that if a few lines were able to be fully loaded, some voltages could be adjusted to their lower restrictions. Although no violation occurred, even a small disturbance, such as a load change, could cause the system to deteriorate into an unstable condition.

As the penetration of wind energy into a power network increases, the influence of wind turbines on the voltage stability becomes more significant. Following a large incorporation of wind power into a grid, severe problems may arise due to both the characteristics of the wind generators and the random nature of wind. Hence, when wind power systems are connected to weak networks, the voltage stability should be considered in addition to the uncertainties of the wind power in their CM frameworks. Voltage stability refers to the ability of a power system to sustain steady levels of voltage for all of the network buses after being subjected to a disturbance [18]. Voltage instability can lead to load shedding, branch trips, or even cascading outages caused by acting protective relays. Voltage collapse is a phenomenon in which a sequence of voltage instability events leads to a blackout. Generally, the closeness to voltage collapse can be used to measure the voltage stability of a power system. In systems with weak connections among areas, congestion problems frequently occur due to either overloading or voltage security requirements. Several techniques have been proposed to solve voltage security problems in CM. In [19], a multi-objective method was presented to maximize both the social benefits of the power market and the distance to the maximum loading point. In this method, which employed a loading margin, the ISO paid more for security enhancement. In [20], a method for ensuring voltage security after the implementation of CM was proposed. However, it did not consider the diverse effects of different loads on the voltage security. In addition, due to the complexities of power-system stability problems, both the solution spaces resulting from inter-area power transfer problems and the system stability boundaries may become complicated. As such, traditional optimization techniques can occasionally fail or encounter numerical difficulties with objective functions. In contrast, both particle swarm optimization (PSO) and differential evolution are powerful population-based searching algorithms [21]. However, transforming a typical single-objective PSO into a multi-objective PSO requires reestablishing both the best local and global individuals to seek a front of optimal solutions.

In this paper, we propose an approach to solving CM problems, focusing in particular on both voltage stability and wind farms in the electricity market. The problem is modeled as a probabilistic voltage-stability constrained optimal power flow (P-VSCOPF) to maximize both the social welfare and the voltage stability margin. A multi-objective particle swarm optimization (MOPSO) algorithm is used to determine the P-VSCOPF solution for mitigating the transmission congestion problem. Since wind speed is a random variable and load forecasting contains uncertainties as well, we apply a probabilistic approach based on Monte Carlo simulations (MCSs). The contingency analysis for line outages is also considered by using the voltage stability index (VSI). The optimal sites of wind farms are determined based on the congestion distribution factor (CDF), and the voltage stability margin (VSM) is considered within the CM framework. Simultaneously, optimal pricing expressions for relieving congestion are derived with linear programming (LP).

2. Uncertainty Analysis and Wind Farm Modeling

2.1. Wind Speed

Wind speed is a random variable, and its fluctuations over a period are expressed by probability distribution functions. In general, wind speed can be described using either a two-parameter Rayleigh or a Weibull distribution [22]. The Rayleigh distribution is the simplest distribution used to describe average wind speed because it has only a single model parameter: b. Its probability distribution and cumulative distribution functions are, respectively,

$$f_{S_w}(S_w; b) = \frac{S_w}{b^2} \exp\left(-\frac{1}{2} \frac{S_w^2}{b^2} \right) \tag{1}$$

and

$$F_{S_w}(S_w; b) = 1 - exp\left(-\frac{1}{2} \frac{S_w^2}{b^2} \right) \tag{2}$$

As a distribution recommended in literature, the Weibull distribution is also widely utilized to both represent wind speed distribution and compute wind energy potential [23]. It is thought to fit the probability distribution for wind speed better than the Rayleigh distribution does because of its more flexible shape granted by its additional parameter. The probability density function of the Weibull distribution is formulated as

$$f_{S_w}(S_w; k, a) = \frac{kS_w^{k-1}}{a^k} \exp\left[-\left(\frac{S_w}{a} \right)^k \right], \ S_w > 0, \quad k, a > 0. \tag{3}$$

The Weibull distribution has a two-parameter function characterized by both a and k. These parameters determine the wind speed necessary for the optimum performance of a wind-energy conversion system. A survey of the literature indicates that the shape parameter of the Weibull distribution for estimating the global wind energy conditions ranges from 1.2 to 2.75. The cumulative distribution function of the Weibull distribution, which gives the probability of the wind speed, is expressed as

$$F_{S_w}(S_w; k, a) = 1 - exp\left[-\left(\frac{S_w}{a}\right)^k\right] \tag{4}$$

2.2. System Load

To reflect a proper probabilistic approach, the uncertainty in the system load must be considered. The pattern of the system load is assessed by accumulating the consumption periodically over daily, weekly, or monthly periods. Uncertainties related to the predicted load data are usually considered by using a normal distribution with a standard deviation. A normal probability distribution function can be expressed as

$$f_L(P_L) = \frac{1}{\sqrt{2\pi\sigma_L^2}} exp\left[-\frac{(P_L - e_L)^2}{2\sigma_L^2}\right] \tag{5}$$

2.3. Wind Turbine Modeling

The power output of a wind turbine is associated with the wind speed, which is converted to electrical power by different types of wind turbine generators. Our model representing a wind turbine's power is given by

$$P_w(S_w) = \begin{cases} 0 & S_w < S_{cut-in} \\ P_{w,rated}\frac{S_w^2 - S_{cut-in}^2}{S_{rated}^2 - S_{cut-in}^2} & S_{cut-in} < S_w < S_{rated} \\ P_{w,rated} & S_{rated} < S_w < S_{cut-out} \\ 0 & S_w > S_{cut-out} \end{cases} \tag{6}$$

Figure 1 shows the representative power curve for a wind turbine. The cut-in wind speed S_{cut-in} is the minimum speed required to generate power. If the wind speed reaches the cut-in value, power is generated by the wind turbine. The generator produces the machine's rated power $P_{w,rated}$ when the wind speed reaches the rated wind speed S_{rated}. As shown in Figure 1, power production is almost constant between S_{rated} and $S_{cut-out}$. At the cut-out wind speed $S_{cut-out}$, power generation is shut down to protect the wind turbine from damage and defects, and the power output becomes zero.

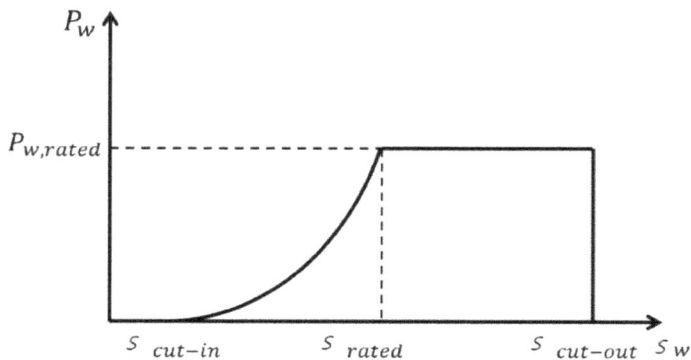

Figure 1. Wind turbine power curve.

Recently, both permanent magnet synchronous generators (PMSGs) and doubly fed induction generators (DFIGs) have begun to be used more widely [24]. PMSG wind turbines have more highly reliable operations, lower maintenance expenses, and smaller weights with simpler structures than DFIG wind turbines do. Accordingly, this paper focuses on PMSG wind turbines. Maximum power point tracking control is usually employed to maximize the turbine energy-conversion efficiency through the regulation of the rotational speeds of variable-speed wind turbines. Depending on the wind's aerodynamic conditions, the power captured by the wind turbine can be maximized versus rotational speed characteristics by adjusting the coefficient G_p, which represents the aerodynamic efficiency of the wind turbine and is a function of the tip speed ratio (TSR). The maximum mechanical output power of the wind turbine is then defined as

$$P_{max} = \frac{1}{2}\rho\pi R^2 S_w^3 G_{p,opt}\left(\beta, \gamma_{w,opt}\right) \tag{7}$$

The optimal TSR of the mechanical output is given as

$$\gamma_{w,opt} = \frac{\omega_{m,opt}R}{S_w} \tag{8}$$

For the PMSG wind turbine modeled in this paper [24], the maximum value of the power coefficient ($G_{p,opt}$ = 0.4412) is obtained according to the optimal value of the TSR ($\gamma_{w,opt}$ = 6.9) calculated using (8). Because wind farms consist of several PMSG wind turbines, the power coefficient of a wind farm located at bus i is the sum of the active powers generated by the wind turbines, given by

$$P_{WF,i} = \sum_{n=1}^{NT_n} P_{WT,n} \tag{9}$$

3. Proposed Congestion Management Approach

3.1. P-VSCOPF

Power systems, including wind farms, are open systems. This means that any external parameter can influence their functionalities, which can lead to rather uncertain systems. The proposed OPF for CM is implemented as a P-VSCOPF, which is a nonlinear, multi-objective optimization problem. A probabilistic method has been developed using MCSs that apply random sampling of the uncertain parameter probability density function to solve the problems. In our study, a multi-objective OPF is formulated as

$$Min \quad -(w_1)(C_D P_D - C_S P_S) - w_2\lambda_c \tag{10}$$

The objective function contains both social welfare and maximum loading margins with weighting factors of $w_1 > 0$ and $w_2 > 0$. Here, we assume that $w_1 = (1-w)$ and $w_2 = w$, (for $0 < w < 1$). The value of the weighting factor is increased so that stability takes precedence over cost. Hence, the system becomes more stable and has higher operating costs.

The equality and inequality constraints for the problem are as follows:
Power flow equations:

$$f(\delta, V, Q_G, P_S, P_D) = 0 \tag{11}$$

$$f(\delta_c, V_c, Q_{G_c}, \lambda_c, P_S, P_D) = 0 \tag{12}$$

Supply and demand bid blocks:

$$P_{S_{min}} \leq P_S \leq P_{S_{max}} \tag{13}$$

$$P_{D_{min}} \leq P_D \leq P_{D_{max}} \tag{14}$$

Generation reactive power limit:

$$Q_{G_{min}} \leq Q_G \leq Q_{G_{max}} \tag{15}$$

$$Q_{G_{min}} \leq Q_{G_c} \leq Q_{G_{max}} \tag{16}$$

Thermal limit:

$$I_{ij}(\delta, V) \leq I_{ij_{max}} \tag{17}$$

$$I_{ji}(\delta, V) \leq I_{ji_{max}} \tag{18}$$

$$I_{ij}(\delta_c, V_c) \leq I_{ij_{max}} \tag{19}$$

$$I_{ji}(\delta_c, V_c) \leq I_{ji_{max}} \tag{20}$$

Voltage security limit:

$$V_{min} \leq V \leq V_{max} \tag{21}$$

$$V_{min} \leq V_c \leq V_{max} \tag{22}$$

Loading margin:

$$\lambda_{c_{min}} \leq \lambda_c \leq \lambda_{c_{max}} \tag{23}$$

Here, the subscript c represents the system's maximum loading condition. Equation (12) is related to a loading parameter, which guarantees that the system network has the required margin of voltage stability. The generation and load in the present state, with the loading parameter λ_c, are expressed as

$$P_G = P_{G_0} + P_S \tag{24}$$

$$P_L = P_{L_0} + P_D \tag{25}$$

$$P_{G_c} = (1 + \lambda_c + k_{G_c})P_G \tag{26}$$

$$P_{L_c} = (1 + \lambda_c)P_L \tag{27}$$

In the proposed P-VSCOPF-based approach, the critical loadability λ_c can be expressed as a measure of system congestion. The loadability is maximized to receive the impact of the voltage stability limit. Therefore, the total maximum loadability (TML) and the available loading capability (ALC) can be defined, respectively, as

$$TML = (1 + \lambda_c)\sum P_{L_i} \tag{28}$$

$$ALC = \lambda_c \sum P_{L_i} = \lambda_c TTL \tag{29}$$

Here, TTL is the total transaction level at the current operating point. In Equation (29), the ALC is computed for the product of λ_c and TTL.

3.2. Stability Margin

To make the system more robust against voltage-related disturbances, a suitable level of voltage stability should always be maintained in the power system. The continuation method is used as a tool for voltage security studies in [25]. In this work, the VSM is applied to measure the voltage stability, and the load is assumed to be of the constant-power type. Figure 2 illustrates the bifurcation curve between the power and the voltage. In the P-V curve, the maximum value of the loading parameter, from the base case up to the saddle node bifurcation (SNB) point, which is regarded as the voltage collapse point, indicates the VSM, which is described as the loading distance between the base case and the SNB point in voltage collapse:

$$VSM = P_{SNB} - P_{base} \tag{30}$$

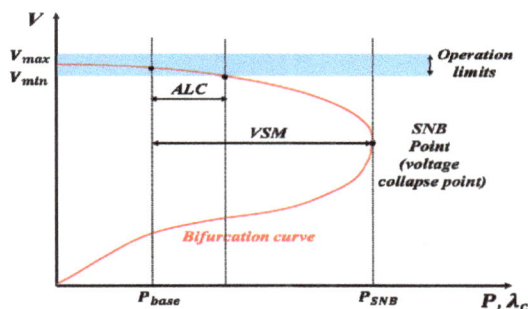

Figure 2. Bifurcation curve for voltage stability analysis.

Equation (30) illustrates that a greater VSM results in a more stable system from a voltage stability viewpoint. On the other hand, a stressed power system typically experiences a low VSM.

3.3. Sensitivity Analysis

3.3.1. VSI Analysis

Steady-state voltage stability analysis includes the determination of an index called the VSI, which is an approximate measure indicating both the most critical bus and the closeness of the system to voltage collapse. To determine the most critical and congested line between buses i and j, contingency ranking is implemented by determining the voltage collapse point from the post-contingency operating conditions. The maximum loading point will move in accordance with variations in both the system topology and the control variables if transmission line outages occur. Since some sensitivity exists between the loading margin and the transmission parameters, a new sensitivity-based branch outage contingency ranking method [26] is used in our study to carry out the contingency selection. This is computed using both the power flow variables and the loading parameter with respect to changes in the transmission line flows. The VSI can be found from the following equation:

$$VSI = \frac{|\Delta P_{ij}| \bullet |P_{ij}|}{|\Delta \lambda_c|} \tag{31}$$

In order to evaluate the weight of each line, a scaling method involving lines with heavy loads is applied. From Equation (31), the contingency line must be the most vulnerable line according to the smallest voltage stability margin. This line is ranked at the top of the list. This technique ranks the contingencies correctly with respect to their impacts on the VSM. The VSI can be used as a real-time operational tool because of its low computational effort requirements. In addition, the VSI is similarly associated with a security cost, which is defined as the amount expended on security improvements to satisfy the $N - 1$ criterion. The security cost contains the additional costs incurred by the system to ensure $N - 1$ security.

3.3.2. CDF

The optimal location of a wind farm can also be obtained through a sensitivity analysis with respect to the power flow injection of any bus n. This analysis introduces information about the variations in power transfer, and therefore, about loading the system with respect to the variations in the power injection of any bus. This study utilizes these sensitivity factors, called CDFs, which are based on the VSI. The derivation of the CDF is given in Appendix A, and the factors are obtained as

$$CDF_n^{i_c j_c} = \frac{\Delta P_{i_c j_c}}{\Delta P_n} \tag{32}$$

3.4. MOPSO Algorithm

The optimization model of P-VSCOPF involves highly complicated implementations of both the objective and constraint equation differentials. To overcome the major restrictions faced by many traditional methods, evolutionary computation algorithms (e.g., PSO or differential evolution) can be used as reliable alternatives in many complex engineering optimization applications [17]. In this paper, the MOPSO based on the Newton–Raphson method is used as the optimization algorithm for the P-VSCOPF problem. The proposed MOPSO is suitable for solving constrained, nonlinear optimization problems. This technique can be managed easily, and it provides good convergence characteristics with high computing efficiencies compared to conventional PSO. Moreover, it obtains good starting values for the initial population before the PSO process begins by using the Newton–Raphson method. Thus, it offers a better performance in finding the optimal Pareto front.

The velocity and position vectors of a particle in an *n*-dimensional space are given by

$$X_a = (x_{a1}, \cdots, x_{an}) \tag{33}$$

$$Z_a = (z_{a1}, \cdots, z_{an}) \tag{34}$$

The best position calculated by a particle is

$$Pbest_a = (x_{a1}{}^{best}, \cdots, x_{an}{}^{best}) \tag{35}$$

The particle among all of the particles in the population that has the best position can be represented as

$$Gbest_g = (x_{g1}{}^{best}, \cdots, x_{gn}{}^{best}) \tag{36}$$

The position and velocity of each particle, updated after $(k + 1)$, steps is calculated by

$$X_a{}^{(k+1)} = X_a^k + Z_a^{k+1} \tag{37}$$

The velocity of the *i*th individual during the $(k + 1)$th iteration can be computed by

$$z_{an}^{t+1} = W z_{an}^t + c_1 r_1 (Pbest_{an}^t - x_{an}^t) + c_2 r_2 (Gbest_{gn}^t - x_{an}^t) \tag{38}$$

Finally, the inertia weight parameter W can be expressed as

$$W = (W_i - W_f) \times \frac{(iter_{max} - iter)}{iter_{max}} + W_f \tag{39}$$

Figure 3 shows the MOPSO process. This application is used to check the feasibility of the non-dominated solutions.

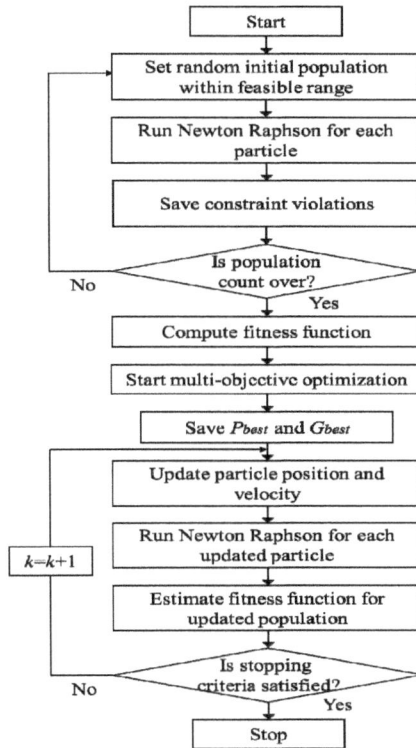

Figure 3. MOPSO procedure.

3.5. Price Analysis

The LMPs are the key factors in both managing the transmission congestion and identifying the spot price. The LMPs are utilized by using a linear programming (LP) approach to compute the power dispatch schedules while simultaneously maximizing the social welfare function. Network congestion causes LMP differences at the buses. These should be minimized to relieve the congestion while also keeping their values as low as possible. LMPs are commonly associated with both the bidding costs and the dual variables (Lagrange multipliers) of the power flow equation [25]. The Lagrange multiplier is the shadow price for the load flow constraints of the OPF, and it describes each congested transmission line on the network. A higher multiplier means, in general, a greater influence of the corresponding congested transmission line on the prices at each location. Using Equations (10)–(27), the expressions for the LMPs are computed to be

$$LMP_{S_i} = \phi_{P_{S_i}} = C_{S_i} + \mu_{P_{S_{max_i}}} - \mu_{P_{S_{min_i}}} - \phi_{cP_{S_i}}(1 + \lambda_c^* + k_{G_c}^*)$$
$$LMP_{D_i} = \phi_{P_{D_i}} = C_{D_i} - \phi_{Q_{D_i}}\tan(\varphi_{D_i}) - \mu_{P_{D_{max_i}}} + \mu_{P_{D_{min_i}}} \tag{40}$$
$$- \phi_{cP_{D_i}}(1 + \lambda_c^*) - \phi_{cQ_{D_i}}(1 + \lambda_c^*)\tan(\varphi_{D_i})$$

Equation (40) contains terms related to the loading parameters, which consider the generation marginal cost of the electricity responsible for both supply and demand, including the system losses of each transmission line. Note that the LMPs have additional terms that rely on the voltage security

29

constraints. Thus, these equations represent the manner in which the voltage stability coordinates the existing prices.

This paper also proposes NCPs to both maintain network security parameter limits and send an appropriate signal based on the price of electricity. Equation (40) can be used to perform a decomposition calculation in order to obtain NCPs that are associated with the transmission line limits. The equation to determine the NCP is given as

$$NCP = \left(\frac{\partial f^T}{\partial X}\right)^{-1} \frac{\partial Z^T}{\partial X} \left(\mu_{max} - \mu_{min}\right) \tag{41}$$

which shows that the NCPs not only involve transmission congestion but also estimate the degree of congestion severity. After clearing the electricity market from the P-VSCOPF solution, our work uses an LMP payment mechanism for the CM as a cost settlement process that is performed according to the market participants' contributions to both the system congestion and the system losses [26]. In doing so, the total ISO payment is computed as the difference between the supplier and consumer payments as

$$Pay_{ISO} = \sum_i C_{S_i} P_{S_i} - \sum_j C_{D_j} P_{D_j} \tag{42}$$

Overall, the proposed approach is implemented sequentially, as shown in Figure 4.

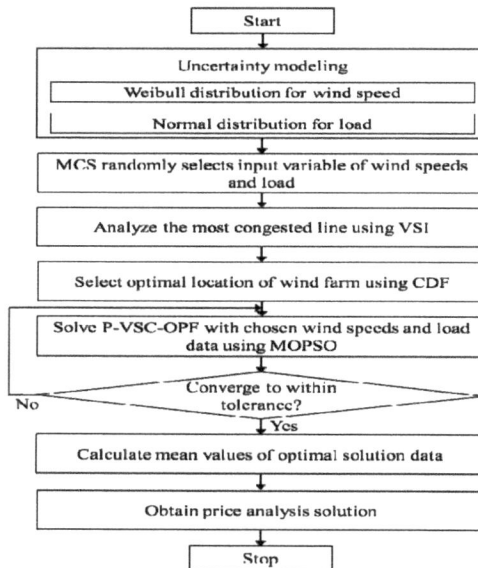

Figure 4. Procedure for proposed approach.

4. Simulation Results

The validity applying of the proposed approach to CM was evaluated using the modified IEEE-24 bus system, as shown in Figure 5. This system consisted of ten generating plants connected by 11 buses, 38 transmission lines, and 17 loads [27], with bus 13 designated as a slack bus. To solve the P-VSCOPF problem, the maximum number of MOPSO algorithm iterations was set to 100. r_1 and r_2 were uniform random factors that were assigned values between 0 and 1 both at each step and for each particle in the swarm in order to add randomness to the velocity update. The inertia weight W was decreased from 0.9 to 0.4 for different iterations. Each of the acceleration coefficients c_1 and c_2 were set equal to 2 in order to achieve a stochastic factor with a mean value of unity. The simulation results were analyzed using the power system analysis toolbox (PSAT) for both MATLAB (version, Manufacturer, City, US State abbrev. if applicable, Country) and GAMS (version, Manufacturer, City, US State abbrev. if applicable, Country) in the optimization package [28].

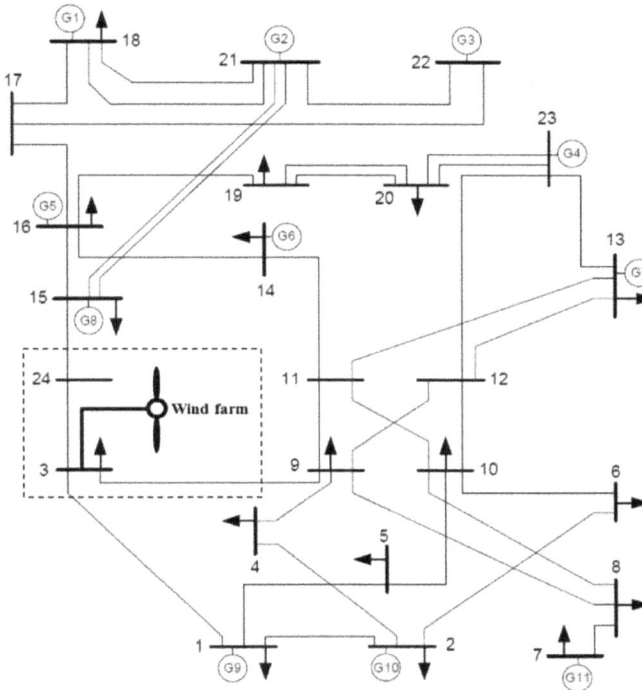

Figure 5. Modified IEEE-24 bus system with a wind farm.

4.1. Probability Distributions of Wind Power and Load

In consideration of the probabilistic variability of both wind speeds and loads, an MCS is conducted to select both the wind power and the system load as input variables. To estimate the Weibull parameters of wind speeds, two years of historical wind speed data were collected from a wind turbine in Seongsan-eup on Jeju Island, South Korea [29]. These data were used as the input for a 2-MW PMSG wind turbine. Our study assumed that wind power generation consumes no fuel. Figure 6 illustrates the cumulative distribution curve of the power output based on a 2-MW wind turbine using an MCS with 1000 samples. The mean value of the wind power output is approximately 1 MW. The normally distributed system load uncertainty is shown in Figure 7. It is considered using the mean value of the period in which congestion occurs more frequently. For statistical purposes, the normal distribution representing the load uncertainty can be obtained with a standard deviation of 5%.

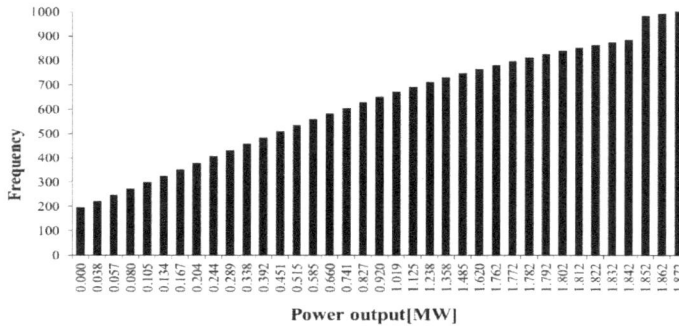

Figure 6. Cumulative distribution curve of wind power.

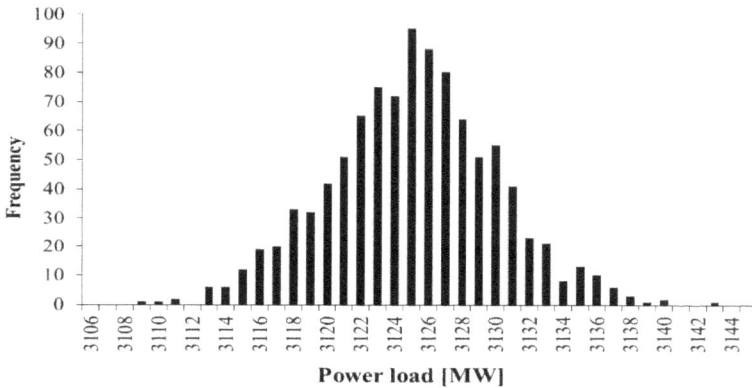

Figure 7. Normally distributed load uncertainty.

4.2. Optimal Locations of Wind Farms

The optimal sites of wind farms with respect to CM were obtained via a CDF based on VSI. In consideration of the contingency analysis, the ten highest transmission-line VSI values were calculated, and the results are shown in Table 1. Note that the line outage between buses 3 and 24 was the most severe because it had the smallest security margin. The critical line, which had the highest security margin value, was considered a potential candidate for CM. These CDFs are obtained based on the VSI data and are given in Table 2. Higher CDFs values indicate that the lines are more critical to power transfer in the transmission lines. Table 2 illustrates that the power injections of bus 3 are the biggest influence on the power flowing through this line. Based on the CDFs results, we find that suitable wind farm locations are typically among the load-side buses. As shown in Figure 5, a wind farm was placed at bus 3, which was the optimal location regarding CM. This wind farm was composed of 300 PMSG wind turbines rated at 2 MW each. The total wind power capacity of 600 MW represented approximately 20% of the total generation capacity.

Table 1. Ranking results and corresponding VSI values.

Rank	Between Buses	VSI
1	3–24	1121.57
2	23–20	735.85
3	17–16	698.50
4	16–14	473.11
5	19–20	475.72
6	19–16	350.73
7	18–17	335.22
8	21–15	304.80
9	10–6	274.51
10	1–2	223.31

Table 2. CDFs for lines between buses 3 and 24.

Bus	CDF	Bus	CDF
1	0.4330	13	−0.0768
2	0.4853	14	−0.3205
3	1.3113	15	−0.7625
4	0.3358	16	−0.634
5	0.0359	17	−0.5492
6	−0.4014	18	−0.5419
7	−0.0498	19	−0.5507
8	0.1511	20	−0.4097
9	0.3992	21	−0.6040
10	0.1580	22	−0.5824
11	−0.0463	23	−0.3308
12	0.0232	24	−1.8239

4.3. Solution Results and Comparison

To verify the effectiveness of the proposed approach, the performances of conventional OPF, VSCOPF, and P-VSCOPF with respect to CM were compared for the modified IEEE-24 bus system. In our study, all three methods were utilized to consider the optimal installation of the wind farm in bus 3, which is obtained through a CDF based on VSI. However, both conventional OPF and VSCOPF, which take into account only the mean values of loads and the wind farm's output, do not consider the probabilistic effects between the loads and the wind farm's variability. Meanwhile, P-VSCOPF consists of two parts: an MCS for selecting both the wind output and load input variables and MOPSO for solving the multi-objective optimization problem. It was implemented based on 1000 MCS trials. Table 3 presents the optimal solution results for P-VSCOPF. Note that these results represent the mean values of 1000 MCS trials based on MOPSO. The power generation (P_G) in bus 3, which was connected to the wind farm, was 295.1 MW. This value was determined by the performance of 1000 MCSs with probabilistic variables taken into account. Figure 8 reveals the estimates of the Pareto optimal front, which relates Pay_{ISO} and the loading parameter λ_c. Note that the Pareto optimal front of the test system exhibits highly nonlinear relationships between the social welfare and the voltage stability margin. A suitable range for the weighting factors is approximated to be 0.4–0.8. In fact, for values smaller than 0.4, system security can almost be ignored, and for values higher than 0.8, market power will happen. On the other hand, with respect to the weighting factors within this range, an importance of between 0.4 and 0.8 was assigned to the voltage security. Therefore, we assume that the results of our work apply to test systems with $w = 0.6$.

Table 3. Optimal solution results for P-VSCOPF.

Bus	LMP ($/MWh)	NCP ($/MWh)	P_G (MW)	P_D (MW)	Pay_{ISO} ($/h)
1	19.4849	0.1934	172	95.04	−1499.55
2	19.5338	0.2067	172	128.04	−858.71
3	18.7920	0.0381	295.1	158.40	2976.65
4	20.1954	0.3120	0	91.48	1847.48
5	19.9423	0.2783	0	62.48	1245.99
6	20.2911	0.3769	0	119.68	2428.43
7	20.2384	0.3134	220.9	161.22	−1208.19
8	20.6143	0.3823	0	150.48	3102.04
9	19.7676	0.2084	0	154.00	3044.22
10	19.8959	0.2633	0	171.60	3414.13
11	19.7557	0.0982	0	0	0
12	19.7065	0.0519	0	0	0
13	19.5299	0.0000	237.8	233.20	−99.62
14	19.5805	0.0460	0	256.08	5014.17
15	18.7755	−0.2834	167	461.64	5532.00
16	18.9073	−0.2417	155	132.00	−434.87
17	18.5877	−0.3417	0	0	0
18	18.7890	−0.3691	400	439.56	−2176.06
19	19.0499	−0.1925	0	238.92	731.42
20	18.9498	−0.2156	0	168.96	4551.40
21	18.3782	−0.3990	400	0	−3201.76
22	17.9248	−0.5160	300	0	−7351.28
23	18.8138	−0.2532	350	0	−12417.08
24	18.5280	−0.3474	0	0	−7040.64

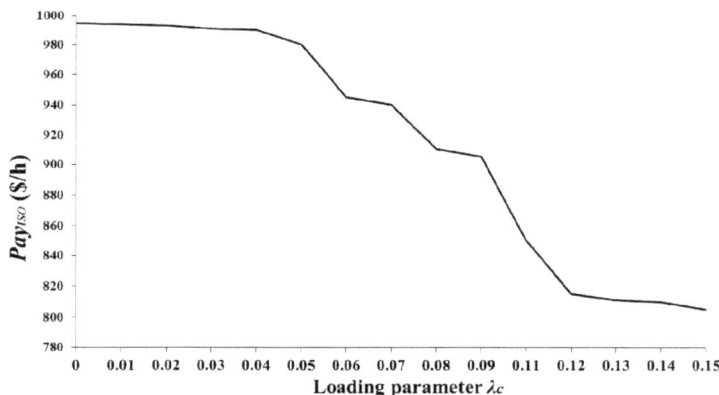

Figure 8. Pareto optimal front for P-VSCOPF.

Figure 9 compares the LMPs and the NCPs. As shown in Figure 9a, for CM, the LMP differences were less with P-VSCOPF than they were with either the conventional OPF or VSCOPF. It is clear that the congestion problems were resolved by minimizing the LMP differences and then maintaining them at the lowest possible levels. As shown in Figure 9b, the range of NCP values was −0.6954 to 0.5962 $/MWh. Here, NCP values at bus 13 (the slack bus) were zero. Lower NCP values indicate lower congestion in the test system. The congestion mitigation of P-VSCOPF was the most effective among those of all of the methods, since the NCP of P-VSCOPF was the lowest. A comparison of the results obtained in each case reveals that the best solution for CM was obtained with P-VSCOPF. Consequently, appropriate realizations of both the voltage stability constraints and the contingency analyses not only result in a better distribution of electricity prices but also reduce the influence of system congestion.

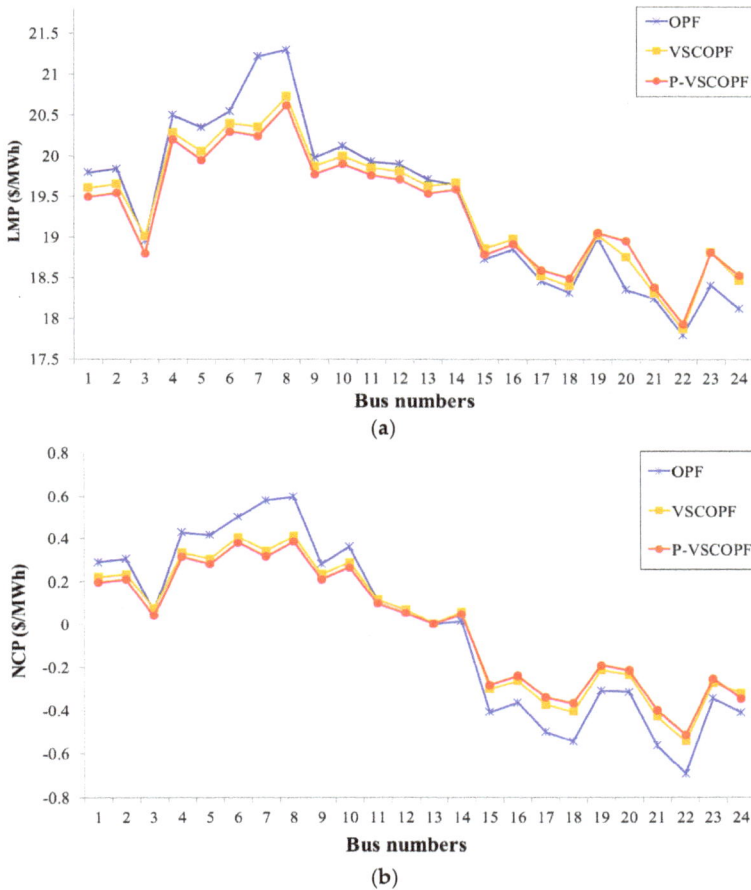

Figure 9. Comparison of optimal pricing for CM. (a) LMPs; (b) NCPs.

Table 4 shows a comparison of the optimal solutions for the three methods. Compared to both the conventional OPF and the VSCOPF, the proposed approach provides higher values for time to live (TTL), TML, and ALC. The use of enhanced LMPs also results in an improved total ISO payment, although the accrued system losses are higher according to the TTL. Meanwhile, the conventional OPF provides the lowest VSM of 364.37 MW after CM has been applied. Despite the fact that the OPF relives congestion, it does not ensure secure operations of the power network. The stability margin could be low, leading to a vulnerable network. When compared to the OPF, the VSCOPF alleviates congestion, ensuring the power system's security in terms of voltage stability. Nevertheless, the stability margin of VSCOPF remains low after congestion relief has taken place. This means that the total ISO payment should increase because of the additional set of system constraints. In addition, neither the conventional OPF nor the VSCOPF considers the probabilistic effects between the congestion and the wind farm's variability, taking only the mean value of the wind farm's output into account. On the other hand, the proposed approach not only provides the system with a larger VSM of 685.43 MW but can also relieve congestion by considering the margin of the system's stability with respect to the uncertainty in the wind farm. The ISO makes selective payments to both participants that can mitigate congestion as well as those that affect the system stability margin significantly.

To evaluate the ongoing security after CM has been executed, the voltage profiles of the buses for the three methods are shown in Figure 10. The horizontal and vertical axes represent the bus numbers and the voltage magnitudes, respectively. A good voltage profile is one of the indications of a more secure power system. The voltages of P-VSCOPF are the highest among all methods for all of the buses. Thus, P-VSCOPF not only provides a greater stability margin but also leads to a better voltage profile than the other OPF methods do. This means that the system is more robust and can maintain stability even under the most severe circumstances.

Table 4. Comparison of optimal solutions of three methods.

Method	TTL (MW)	System Losses (MW)	TML (MW)	ALC (MW)	Pay_{ISO} ($/h)	VSM (MW)
Conventional OPF	3056.97	52.63	3246.19	189.22	1049.95	364.37
VSCOPF	3139.86	42.39	3555.11	385.25	820.13	640.87
P-VSCOPF	3222.77	41.92	3638.04	415.36	812.26	685.43

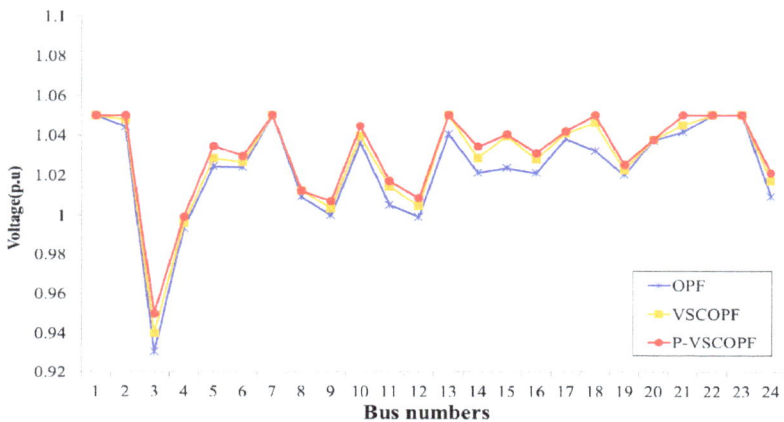

Figure 10. Voltage profiles of three methods.

5. Conclusions

In this paper, a probabilistic, multi-objective approach for CM that considers both voltage stability and wind energy was proposed. P-VSCOPF was formulated as a multi-objective problem not only for maximizing social welfare but also for improving the voltage stability margin. It can be used to assist both ISOs and planners in visualizing the nonlinear relationships between power transfer levels and voltage stability margins via the MOPSO algorithm. MCSs were applied to select the input variables of both wind speed and load by using the probability distribution functions. The problem was solved using 1000 trials of an MCS. The VSI was calculated to increase the accuracy of the sorting and ranking technique in the contingency analysis. The optimal location of a wind farm, with a special emphasis on CM, was also determined using the CDF. The influence of the wind output variations on the congestion price, which is related to both the LMPs and the NCPs, was analyzed. The VSM, ALC, and TML were employed as measures of voltage stability. To illustrate its effectiveness, the proposed approach was both demonstrated and compared to the conventional OPF and VSCOPF using the modified IEEE-24 bus system with wind farms. The simulation results revealed that the P-VSCOPF achieved discernible advantages over all of the other OPF methods. It was able to not only minimize the LMP differences and lower the NCP for CM but also increase the VSM. Therefore, the information provided by the proposed approach can be useful in both the planning and operation of various power systems.

Acknowledgments: This research was supported by Korea Electric Power Corporation through Korea Electrical Engineering & Science Research Institute. (grant number: R15XA03-55).

Author Contributions: Jin-Woo Choi proposed the content of this paper and Mun-Kyeom Kim coordinated the proposed manuscript approach.

Conflicts of Interest: The authors declare no conflict of interest.

Nomenclature

Constants

ρ	Air density (=1.205 kg/m^2)
$\gamma_{w,opt}$	Optimal tip speed ratio
$G_{p,opt}$	Optimal power coefficient of wind turbine
R	Rotator radius
$\omega_{m,opt}$	Optimal rotational speed
S_{cut-in}	Cut-in speed of wind turbine
S_{rated}	Rated speed of wind turbine
$S_{cut-out}$	Cut-out speed of wind turbine
$P_{\max}, P_{\min}$	Maximum and minimum limits of possible power
$\lambda_{c_{\max}}, \lambda_{c_{\min}}$	Maximum and minimum limits of loading margin
$P_{S_{\max}}, P_{S_{\min}}$	Maximum and minimum limits of supply bid
$P_{D_{\max}}, P_{D_{\min}}$	Maximum and minimum limits of demand bid
$Q_{G_{\max}}, Q_{G_{\min}}$	Maximum and minimum limits of reactive power
$I_{ij_{\max}}$	Maximum limit of line currents between nodes i and j
$V_{\max}, V_{\min}$	Maximum and minimum limits of voltage magnitude
$\mu_{\max}, \mu_{\min}$	Maximum and minimum limits of shadow price
T	Scheduling time (e.g., 24 h)
c_1, c_2	Acceleration coefficients

Variables

β	Blade pitch angle
S_w	Wind speed
b	Scale parameter of Rayleigh distribution
k, a	Shape and scale parameter of Weibull distribution
σ_L	Standard deviation of load
e_L	Expected value of probability variable
w	Weighting factor
$P_{w,rated}$	Rated power of wind turbine
$P_{WF,i}$	Active power output of wind farm at bus i
$P_{WF,n}$	Active power of nth wind turbine
ΔP_n	Variation in active power of nth bus
P_G, P_L	Power outputs of generator and load, respectively
P_{G_o}, P_{L_o}	Current power outputs of generator and, respectively
P_S, P_D	Supply and demand bid volumes (in MW), respectively
C_S, C_D	Bid prices for supply and demand (in \$/MWh)
Q_G	Generator reactive power
$\Delta P_{i_c j_c}$	Variation of power flow in critical condition
P_{ij}, P_{ji}	Power flowing through lines in both directions
V_i, V_j	Voltage magnitude at buses i and j, respectively
I_{ij}, I_{ji}	Line currents in both directions
P_{SNB}	Power output of SNB point for voltage collapse
P_{base}	Power output at base operating point
δ_i, δ_j	Voltage angle at buses i and j, respectively
θ_{ij}, Y_{ij}	Angle and magnitude of ijth element of Y_{bus}, respectively

λ_c	Loading parameter under critical conditions
k_{G_c}	Loss distribution factor
φ	Lagrange multiplier
μ	Dual variable
φ_{D_i}	Constant load demand power factor angle
X_a^k, Z_a^k	Position and velocity of ith particle at iteration k, respectively
W_i, W_f	Initial and final values of inertia weight

Numbers and Sets

J_r	Reduced Jacobian matrix
NT_n	Number of wind turbines
k	Number of iterations
$iter_{max}$	Maximum number of allowed iterations
r_1, r_2	Random numbers between 0 and 1

Appendix A. Derivation of CDF

The real power flow on line k connected between bus i and bus j is formulated as [30]:

$$P_{ij} = |V_i||V_j||Y_{ij}|cos(\theta_{ij} - \delta_i + \delta_j) - V_i^2 Y_{ij} \, cos \, \theta_{ij}. \tag{A1}$$

Applying the approximation of a Taylor series expansion and ignoring the effects of the remaining higher-order terms, Equation (A1) gives

$$\Delta P_{ij} = \frac{\partial P_{ij}}{\partial \delta_i}\Delta\delta_i + \frac{\partial P_{ij}}{\partial \delta_j}\Delta\delta_j + \frac{\partial P_{ij}}{\partial V_i}\Delta V_i + \frac{\partial P_{ij}}{\partial \delta_j}. \tag{A2}$$

Equation (A2) can be rewritten as

$$\Delta P_{ij} = a_{ij}\Delta\delta_i + b_{ij}\Delta\delta_j + c_{ij}\Delta V_i + d_{ij}\Delta V_j. \tag{A3}$$

The coefficients in Equation (A3) are formulated from the partial derivatives of the real power injection corresponding to the variables δ and V and given as

$$a_{ij} = V_i V_j Y_{ij} \, sin(\theta_{ij} + \delta_j - \delta_i), \tag{A4}$$

$$b_{ij} = -V_i V_j Y_{ij} \, sin(\theta_{ij} + \delta_j - \delta_i), \tag{A5}$$

$$c_{ij} = V_j Y_{ij} \, cos(\theta_{ij} + \delta_j - \delta_i) - 2V_i Y_{ij} \, cos \, \theta_{ij}, \tag{A6}$$

$$d_{ij} = V_i Y_{ij} \, cos(\theta_{ij} + \delta_j - \delta_i). \tag{A7}$$

The Newton–Raphson Jacobian relationship is considered in determining the CDFs as

$$\begin{bmatrix} \Delta P \\ \Delta Q \end{bmatrix} = [J] \begin{bmatrix} \Delta\delta \\ \Delta V \end{bmatrix} = \begin{bmatrix} J_{11} & J_{12} \\ J_{21} & J_{22} \end{bmatrix} \begin{bmatrix} \Delta\delta \\ \Delta V \end{bmatrix} \tag{A8}$$

Taking the coupling between ΔP–$\Delta\delta$ and ΔQ–$\Delta\delta$ into consideration and assuming that the reactive power flows are constant (i.e., $\Delta Q = 0$), the power injections variations can be expressed by

$$\Delta P = J_{11}\Delta\delta + J_{12}\Delta V, \tag{A9}$$

$$0 = J_{21}\Delta\delta + J_{22}\Delta V. \tag{A10}$$

Equations (A9) and (A10) then become

$$\Delta P = J_{11}\Delta\delta - J_{12}J_{22}^{-1}J_{21}\Delta\delta = J_r\Delta\delta. \tag{A11}$$

From Equation (A11), the value of the voltage angle variation is given by

$$\Delta\delta = [J_r]^{-1}\Delta P. \tag{A12}$$

From Equation (A10), the voltage variation with respect to the variations in power can be written as

$$\Delta V = J_{22}^{-1}J_{21}\Delta\delta = J_{22}^{-1}J_{21}[J_r]^{-1}\Delta P. \tag{A13}$$

Equations (A12) and (A13) are formulated as

$$\Delta\delta_i = \sum_{l=1}^{n} m_{il}\Delta P_l \quad i = 1,\ 2,\ ...,n,\ \ i \neq s, \tag{A14}$$

$$\Delta V_i = \sum_{l=1}^{n} m_{ilv}\Delta P_l \quad i = 1,\ 2,\ ...,n,\ \ i \neq s. \tag{A15}$$

Substituting Equations (A14) and (A15) into Equation (8), the variation in real power flow becomes

$$\Delta P_{ij} = a_{ij}\sum_{l=1}^{n} m_{il}\Delta P_l + b_{ij}\sum_{l=1}^{n} m_{jl}\Delta P_l + c_{ij}\sum_{l=1}^{n} m_{ilv}\Delta P_l + d_{ij}\sum_{l=1}^{n} m_{jlv}\Delta P_l. \tag{A16}$$

Equation (A16) can be rewritten as

$$\Delta P_{ij} = \begin{array}{l} (a_{ij}m_{i1} + b_{ij}m_{j1} + c_{ij}m_{i1v} + d_{ij}m_{j1v})\Delta P_1 \\ +...(a_{ij}m_{in} + b_{ij}m_{jn} + c_{ij}m_{inv} + d_{ij}m_{jnv})\Delta P_n \end{array}. \tag{A17}$$

Thus, the CDF corresponding to both the *n*th bus and the line *ij*, connecting buses *i* and *j*, can be obtained as

$$\Delta P_{ij} = CDF_1^{ij}\Delta P_1 + CDF_2^{ij}\Delta P_2 + ... + CDF_n^{ij}\Delta P_n, \tag{A18}$$

$$CDF_n^{ij} = a_{ij}m_{in} + b_{ij}m_{jn} + c_{ij}m_{inv} + d_{ij}m_{jnv}. \tag{A19}$$

References

1. Kumar, A.; Srivastava, S.C.; Singh, S.N. Congestion management in competitive power market: A bibliographical survey. *Int. J. Electr. Power Energy Syst.* **2005**, *76*, 153–164. [CrossRef]
2. Méndez, R.; Rudnick, H. Congestion management and transmission rights in centralized electric markets. *IEEE Trans. Power Syst.* **2004**, *19*, 889–896. [CrossRef]
3. Pillay, A.; Karthikeyan, S.P.; Kothari, D. Congestion management in power systems—A review. *Int. J. Electr. Power Energy Syst.* **2015**, *70*, 83–90. [CrossRef]
4. Hong, Y.Y.; Wu, C.P. Day-ahead electricity price forecasting using a hybrid principal component analysis network. *Energies* **2012**, *5*, 4711–4723. [CrossRef]
5. Wang, Q.; Zhang, G.; McCally, J.D.; Zheng, T.; Litvinov, E. Risk-Based Locational Marginal Pricing and Congestion Management. *IEEE Trans. Power Syst.* **2014**, *29*, 2518–2528. [CrossRef]
6. Pandey, S.N.; Tapaswi, S.; Srivastava, L. Growing RBFNN-based soft computing approach for congestion management. *Neural Comput. Appl.* **2009**, *18*, 945–955. [CrossRef]
7. Xie, J.; Wang, L.; Bian, Q.; Zhang, X.; Zeng, D.; Wang, K. Optimal Available Transfer Capability Assessment Strategy for Wind Integrated Transmission Systems Considering Uncertainty of Wind Power Probability Distribution. *Energies* **2016**, *9*, 704. [CrossRef]
8. Tan, A.; Lin, X.; Sun, J.; Lyu, R.; Li, Z.; Peng, L.; Khalid, M.S. A Novel DFIG Damping Control for Power System with High Wind Power Penetration. *Energies* **2016**, *9*, 521. [CrossRef]
9. De Quevedo, P.M.; Contreras, J. Optimal Placement of Energy Storage and Wind Power under Uncertainty. *Energies* **2016**, *9*, 528. [CrossRef]
10. Sood, Y.R.; Singh, R. Optimal model of congestion management in deregulated environment of power sector with promotion of renewable energy sources. *Renew. Energy* **2010**, *35*, 1828–1836. [CrossRef]

11. Deb, S.; Gope, S.; Goswami, A.K. Congestion management considering wind energy sources using evolutionary algorithm. *Electr. Power Compon. Syst.* **2015**, *43*, 723–732. [CrossRef]

12. Ahmadi, H.; Lesani, H. Transmission congestion management through LMP difference minimization: A renewable energy placement case study. *Arab. J. Sci. Eng.* **2014**, *39*, 1963–1969. [CrossRef]

13. Morgan, E.C.; Lackner, M.; Vogel, R.M.; Baise, L.G. Probability distributions for offshore wind speeds. *Energy Convers. Manag.* **2011**, *52*, 15–26. [CrossRef]

14. Aguado, J.A.; Quintana, V.H.; Madrigal, M.; Rosehar, W.D. Coordinated spot market for congestion management of inter-regional electricity markets. *IEEE Trans. Power Syst.* **2004**, *19*, 180–187. [CrossRef]

15. Saini, A.; Saxena, A.K. Optimal power flow based congestion management methods for competitive electricity market. *Int. J. Eng. Comput. Electr. Eng.* **2010**, *2*, 73–80. [CrossRef]

16. Talukdarar, B.K.; Sinhaa, A.K.; Mukhopadhyaya, S.; Bose, A. A computationally simple method for cost-efficient generation rescheduling and load shedding for congestion management. *Int. J. Electr. Power Energy Syst.* **2005**, *77*, 379–388. [CrossRef]

17. Su, H.Y.; Hsu, Y.L.; Chen, Y.C. PSO-based voltage control strategy for loadability enhancement in smart power grids. *Appl. Sci.* **2016**, *6*, 449. [CrossRef]

18. Esmaili, M.; Shayanfar, H.A.; Amjady, N. Congestion management considering voltage security of power systems. *Energy Convers. Manag.* **2009**, *50*, 2562–2569. [CrossRef]

19. Milano, F.; Cañizares, C.A.; Invernizzi, M. Multiobjective optimization for pricing system security in electricity markets. *IEEE Trans. Power Syst.* **2003**, *18*, 596–604. [CrossRef]

20. Conejo, A.J.; Milano, F.; García-Bertrand, R. Congestion management ensuring voltage stability. *IEEE Trans. Power Syst.* **2006**, *21*, 357–364. [CrossRef]

21. Pandit, M.; Chaudhary, V.; Dubey, H.M.; Panigrahi, B.K. Multi-period wind integrated optimal dispatch using series PSO-DE with time-varying Gaussian membership function based fuzzy selection. *Int. J. Electr. Power Energy Syst.* **2015**, *73*, 259–272. [CrossRef]

22. Manwell, J.F.; McGowan, J.G.; Rogers, A. *Wind Energy Explained: Theory Design and Application*; John Wiley & Sons: Chichester, UK, 2002.

23. Ramirez, P.; Carta, J.A. Influence of the data sampling interval in the estimation of the parameters of the Weibull wind speed probability density distribution: A case study. *Energy Convers. Manag.* **2005**, *46*, 2419–2438. [CrossRef]

24. Choi, J.W.; Heo, S.Y.; Kim, M.K. Hybrid operation strategy of wind energy storage system for power grid frequency regulation. *IET Gener. Trans. Distrib.* **2016**, *10*, 736–749. [CrossRef]

25. Kim, M.K.; Hur, D. An optimal pricing scheme in electricity markets by parallelizing security constrained optimal power flow based market-clearing model. *Int. J. Electr. Power Energy Syst.* **2013**, *48*, 161–171. [CrossRef]

26. Kim, M.K.; Park, J.K.; Nam, Y.W. Market clearing for pricing system security based on voltage stability criteria. *Energy* **2011**, *36*, 1255–1264. [CrossRef]

27. Kim, S.S.; Kim, M.K.; Park, J.K. Consideration of Multiple Uncertainties for Evaluation of Available Transfer Capability using fuzzy continuation power flow. *Int. J. Electr. Power Energy Syst.* **2008**, *30*, 581–593. [CrossRef]

28. Ferris, M.C. *MATLAB and GAMS: Interfacing Optimization and Visualization Software*; Comuter Sciences Department, University of Wisconsin-Madison: Madison, WI, USA, 2005. Available online: http://www.cs.wisc.edu/math-prog/matlab.html (accessed on 7 April 2017).

29. Min, C.G.; Hur, D.; Park, J.K. Economic Evaluation of Offshore Wind Farm in Korea. *J. Electr. Eng. Technol.* **2014**, *63*, 1192–1198. [CrossRef]

30. Kumar, A.; Srivastava, S.C.; Singh, S.N. A zonal congestion management approach using ac transmission congestion distribution factors. *Electr. Power Syst. Res.* **2014**, *72*, 85–93. [CrossRef]

applied
sciences

MDPI

Article

Coordination of EVs Participation for Load Frequency Control in Isolated Microgrids

Mostafa Vahedipour-Dahraie [1], Homa Rashidizaheh-Kermani [1], Hamid Reza Najafi [1,*], Amjad Anvari-Moghaddam [2] and Josep M. Guerrero [2]

[1] Department of Electrical & Computer Engineering, University of Birjand, Birjand 9856, Iran; vahedipour_m@birjand.ac.ir (M.V.-D.); rashidi_homa@birjand.ac.ir (H.R.-K.)

[2] Department of Energy Technology, Aalborg University, 9220 Aalborg East, Denmark; aam@et.aau.dk (A.A.-M.); joz@et.aau.dk (J.M.G.)

* Correspondence: h.r.najafi@birjand.ac.ir; Tel.: +98-56-3220-2049

Academic Editor: José L. Bernal-Agustín
Received: 30 April 2017; Accepted: 19 May 2017; Published: 24 May 2017

Abstract: Increasing the penetration levels of renewable energy sources (RESs) in microgrids (MGs) may lead to frequency instability issues due to intermittent nature of RESs and low inertia of MG generating units. On the other hand, presence of electric vehicles (EVs), as new high-electricity-consuming appliances, can be a good opportunity to contribute in mitigating the frequency deviations and help the system stability. This paper proposes an optimal charging/discharging scheduling of EVs with the goal of improving frequency stability of MG during autonomous operating condition. To this end, an efficient approach is applied to reschedule the generating units considering the EVs owners' behaviors. An EV power controller (EVPC) is also designed to determine charge and discharge process of EVs based on the forecasted day-ahead load and renewable generation profiles. The performance of the proposed strategy is tested in different operating scenarios and compared to those from non-optimized methodologies. Numerical simulations indicate that the MG performance improves considerably in terms of economy and stability using the proposed strategy.

Keywords: microgrid (MG); renewable energy sources (RESs); electric vehicle (EV); frequency stability; energy management strategy (EMS)

1. Introduction

Microgrids (MGs) are a part of distribution systems that include several means of distributed generation (DG), renewable energy sources (RESs), storage devices and controllable loads and have the capability to operate either in connected or isolated mode [1,2]. During islanded operation, due to low inertia of MG and intermittent nature of RESs such as wind and solar, there might be some frequency deviations beyond the acceptable range. Thus, an islanded MG requires specific primary and secondary frequency control schemes, in order to maintain power balance between generation and load and restore frequency to the nominal value [3]. Moreover, MG requires sufficient spinning reserve provided by DG units or energy storage systems (ESSs) to keep power balance during islanded operation [4]. With increasing penetration level of EVs and considering that they are available most of the times in a day, they can play the role of ESSs in a way to alter their energy consumption/production level under the vehicle-to-grid (V2G) concept and exchange the power with the grid [5]. Thus, with V2G capability, EVs can provide ancillary services for the grid, such as frequency regulation [5], load levelling [6,7] and spinning reserve [8]. With the application of a well-designed energy management system (EMS), EVs can act as an effective solution to compensate the uncertain behaviour of RESs. On the other hand, EVs without any proper management strategy could cause a number of issues such as energy losses, overloads, and voltage and frequency fluctuations [9]. Different methods have been used in recent literature to manage EVs' charging and discharging process to facilitate ancillary services in the presence of RESs [10,11]. The role of EVs in primary frequency response and in the presence of renewable energies was investigated in [5] for the Great Britain power system. By considering three EV charging strategies, it was shown that a proper EV charging strategy is effective in primary frequency response and can stabilize the grid frequency when is needed. In [12], with regard to the randomness of renewable energy generation, the operation of EVs in the MG was scheduled to minimize cost of charging. Moreover, economic incentives for EV owners to compensate the wind forecast uncertainties were provided in [13]. Coordination of EVs and minimization of the penalty cost associated with wind power imbalances was studied in [14]. Authors in [15] presented an aggregated primary frequency control model, where a participation factor, based on the state-of-charge (SOC), was used to determine the droop characteristic. It was investigated that EVs can effectively improve the system frequency response due to their ability to participate in primary reserves. The work in [16] performed a comparative study in order to evaluate benefits of EVs providing primary frequency control in an islanded system with high penetration of RESs. A control strategy was also presented in [17] in order to provide active participation of EVs for load frequency response purposes. In the same study, the SOC of EVs was managed by using a smart charging strategy in order to obtain a scheduled charging level requested by an EV owner. In a similar manner, Ref. [18] provided frequency regulation to the power grid using EVs with an effective pricing policy, and Ref. [19] presented a frequency control method considering both EVs and controllable loads.

When an MG enters an isolated mode due to loss of the main grid or a blackout/fault, it necessitates appropriate control and management schemes such that both the economical and stability targets are achieved. In this regard, this paper proposes an EV energy management scheme for MG autonomous operation based on local frequency measurements such that EVs will contribute. The participation of EVs will contribute to the frequency stability of the MG in off-grid times. The proposed control strategy is intended to manage EVs' charging and discharging process considering cost signal and the demand not supplied by renewable resources (DNS_{Ren}) in order to keep the power balance within the MG. The major contributions of this paper are summarized as follows:

- Developing a stability margin index considering variability of load and renewable resources generation to attain the electricity cost signal to manage EVs' charging and discharging process.
- Application of a new EV power controlling (EVPC) scheme to improve frequency stability of an islanded MG.
- Maximize the MG operator's profit and minimize total emission of generating units.

The remainder of this paper is organized as follows. The EV power controlling (EVPC) scheme is discussed in Section 2. The problem of techno-economic optimization of the MG is formulated in Section 3. The simulation results and discussion are expressed in Section 4 and the conclusion is drawn in Section 5.

2. EV Power Controlling (EVPC) Scheme

2.1. MG Energy Management and Control

The stochastic behavior of RESs might cause inevitable concerns for the reliable operation of an islanded-MG. Any power fluctuation in such energy resources may lead to imbalance between load and generation, and, as a result, the frequency may deviate from its nominal value. On the other hand, an appropriate EV energy management strategy can respond to system frequency deviation and thus provide primary frequency regulation. EVs can supply (absorb) energy to (from) the network considering two possible operating modes: grid to vehicle (G2V) mode (charge and absorb power) and V2G mode (discharge and inject power to the network). Thus, from the MG operator's point of view, EVs can act both as load and generation. EVs participate in frequency control to charge in low load or high generation hours and discharge during high load or low renewable generation hours. In order to measure the power mismatch between demand and renewable resources generation, an indicator is introduced here as demand not supplied by renewable sources (DNS_{Ren}) and defined as below:

$$DNS_{Ren}(t) = D(t) - P_{Ren}(t) \quad \forall t \in T, \tag{1}$$

where $D(t)$ and $P_{Ren}(t)$ are total demand and generation of renewable resources in period t, respectively. In an isolated MG without energy storage, DNS_{Ren} should be supplied by dispatchable generators (DGs). The difference between the maximum installed capacity of DG units and DNS_{Ren} stands for the spinning reserve capacity of MG. Here, the normalized value of this capacity is defined as stability margin index (*SMI*):

$$SMI(t) = 1 - \frac{DNS_{Ren}(t)}{\sum\limits_{i \in NG} P_i^{max}} \quad \forall t. \tag{2}$$

The SMI index value at each time period depends on the installed capacity of DG units, output power of renewable resources and customers' demand. As an illustrative example, the variation of SMI during 24 h of a given day is shown in Figure 1. When the total demand is supplied by renewable resources, SMI value is 1 and it is zero if DNS_{Ren} is equal to the total installed capacity of DGs. If the generated power of RESs is more than the total demand, then SMI will be more than 1 (e.g., point P_1 in Figure 1). In this case, as it can be observed, the system frequency is unstable, and, in order to keep frequency within its nominal value, dump load (DL), which is comprised of a set of three-phase resistors connected in series, is used in order to be activated to absorb that excess power. Moreover, if a portion of load is supplied by RESs, the SMI index takes a value between 0 and 1. In this case, the system frequency remains within its nominal values, but its variation differs in various points of SMI. For example, as it can be observed from Figure 1, the frequency at points P_2 and P_3 has different variations but is still in the accepted range. Furthermore, when SMI value is less than zero (i.e., point P_4 in the system frequency drops, thus it is required to inject power to the network (for example, by discharging of EVs)). In the next step, SMI is applied to obtain electricity buying and selling prices for EVs' energy management.

The cost signal should follow the SMI index in such a way to encourage EV owners to charge or discharge their vehicles for keeping the power balance of the system and improving the frequency. Thus, cost signal should track the SMI trend with considering the range of electricity buying price (EBP) and electricity selling price (ESP) obtained from electricity market. On the other hand, the maximum (minimum) value of electricity price represents the highest (lowest) value of SMI. Other prices are fitted between these two limitations based on the values of SMI with the application of numerical analysis.

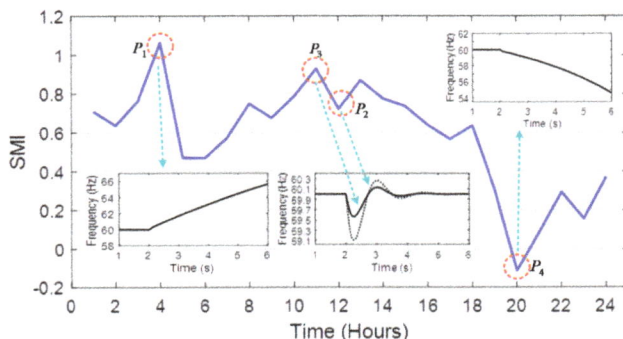

Figure 1. A typical stability margin index (SMI).

2.2. EVPC Structure

In calculating the payments to (by from) the MG operator when EVs are charged (discharged), EBP (ESP) value is introduced to reflect both the wholesale electricity prices and domestic tariffs [20]. Generally, once an EV is connected to the MG, it will be charged and thus has to pay for the consumed energy based on the EBP. However, when it discharges, it receives payments based on the ESP for providing the service. The high level of ESP is considered as a high electricity selling price (HESP), and, here, it is set at 85% of the maximal ESP. Moreover, the high level of EBP is considered as high electricity buying price (HEBP), and it is set at 60% of the maximal EBP. In the islanding operation mode of MG, the major goal is to keep the system power balance and, consequently, to limit frequency variations within the allowed range. Considering the same objective, this paper presents an EVPC scheme as an energy management strategy in order to manage the participation of EVs in charging/discharging process with considering the intermittent behavior of renewable resources. Figure 2 shows the algorithm of an EVPC scheme for optimal scheduling of the MG, and it includes three stages. In the first stage, the difference between day-ahead load and the forecasted power of renewable resources would be calculated to obtain stability margin index and the electricity cost signal that are utilized for managing the power of EVs. This stage includes the following parts:

- Forecasted load and output powers of RESs (wind and solar); it is assumed that the day-ahead load demand and renewable generations are determined.
- Index calculation part; in this block, SMI is calculated using Labels (1) and (2).
- Price calculation part; the electricity prices are obtained based on SMI index.

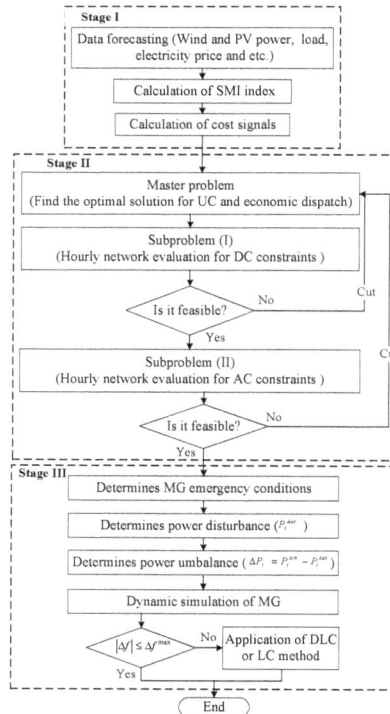

Figure 2. Algorithm of the electric vehicle power controller (EVPC) scheme for optimal scheduling of the microgrid (MG).

In the second stage, an optimal scheduling of the generating units is done to match the demand for the scheduling horizon properly through a unit commitment algorithm and optimal power flow procedure by considering system's objectives and constraints (technical and security).

In the third stage, the MG emergency operation is determined and the frequency behavior is evaluated. Since, the MG is operating in islanded mode, the power unbalance (P_t^{dist}) will result from the changes in loads or generation. The emergency active power of unit i at the certain time t ($dP_{i,t}$), is defined as:

$$\begin{cases} dP_{i,t} = P_t^{dist} \cdot \dfrac{R_{i,t}}{R_t} \\ \sum\limits_{i=1}^{N_G} dP_{i,t} = P_t^{dist} \end{cases} \forall t, \tag{3}$$

where $R_{i,t}$ is the reserve capacity of unit i at time period t, and R_t is the total available generation reserve at time period t.

The demand load, EV power and the generation emergency dispatch are applied as inputs to the MG dynamic model, in order to evaluate the energy balance within the MG for a given period and the expected frequency deviation in the event of a disturbance. Based on the dynamic model results, if the MG does not have enough reserve capacity, it is necessary to exploit emergency load curtailment (LC). Moreover, DL is applied when RESs' generated power is high and a portion of it is not consumed.

2.3. Coordination of EVs Operation with EVPC

Different types of charge and discharge of EVs are considered based on electricity cost signal and their *SOC* when plugged in: charging with high current (I_H), charging/discharging with medium current (I_M) and low current (I_L). The flowchart of delivered power of EV k is depicted in Figure 3. The transacted power of each EV would be obtained based on its initial conditions such as *SOC* and the calculated EBP and ESP based on SMI index. The relationship between *SOC* of EV k and the charge/discharge current is obtained from (4) [9]:

$$SOC_{k,t} = 1 - \frac{I_{k,t} \times \Delta t_{plug}}{3600 C_k^a}, \tag{4}$$

where $I_{k,t}$, Δt_{plug} and C_k^a are the current, plugged-in time (in seconds) and the available capacity of the battery. Since most EV companies use lead acid batteries for their vehicles, here, this kind of technology is also adopted. The voltage against released capacity at different discharge/charge currents for a lead-acid battery is extracted from [20].

In addition, in idle or driving mode, there is no power transaction between EV and the MG, but the stored energy might decrease depending on the length of its daily travel (L_k) and its energy consumption per km (r_k). Each EV comes back to the parking lot after driving L_k km and its *SOC* at the entrance of parking lot (SOC_k^{ent}) is obtained as following:

$$SOC_k^{ent} = SOC_k^{int} - L_k \times r_k, \tag{5}$$

where SOC_k^{int} is the SOC at the start of a day trip.

The value of EV SOC at time t is obtained with respect to the amount of its initial SOC, charge/discharge energy (when it is connected to the grid) and the energy decreased due to its travelling. On the other hand, SMI index at time t would be calculated from the previously-mentioned equations. With considering both SOC and SMI indexes, the charge/discharge process would be determined. In other words, if EV is connected, the value of current that EV absorbs (injects) from (to) the network would be found. Then, the voltage would be obtained based on the voltage-current characteristic of the battery. Thus, with the obtained voltage and current, the transacted power between EV and the network would be captured. This process repeats at each time t.

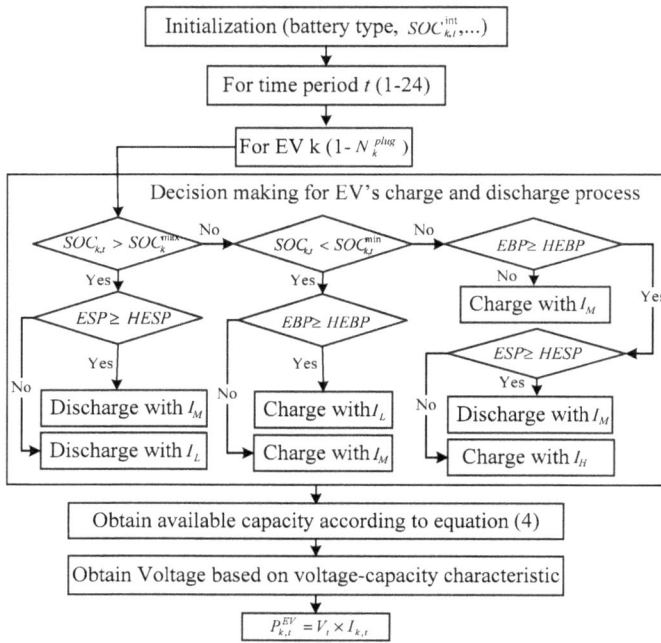

Figure 3. Flowchart of delivered power of electric vehicle (EV) k.

3. Problem Formulation

In order to investigate techno-economic valuation and optimization of the proposed strategy in the MG, a multi-objective optimization problem is formulated with several objectives.

3.1. Objective Function

In the proposed multi-objective problem, three objectives are considered as the MG operator's profit (F_1) maximization, and minimization of the total emission (F_2) as well as the cost corresponding to frequency deviation of the MG (F_3). The total MG operator's profit is formulated as:

$$
\begin{aligned}
F_1 &= \sum_{t=1}^{T}\sum_{j=1}^{N_J} C_{j,t} \cdot D_{j,t} \\
&\quad - \sum_{t=1}^{T}\sum_{i=1}^{N_G} [(A_i \cdot u_{i,t} + B_i \cdot P_{i,t}) + SUC_i \cdot y_{i,t} + SDC_i \cdot z_{i,t}] \\
&\quad - \sum_{t=1}^{T}\sum_{w=1}^{N_W} C_{w,t} \cdot P_{w,t} - \sum_{t=1}^{T}\sum_{p=1}^{N_P} C_{p,t} \cdot P_{p,t} \\
&\quad + \sum_{t=1}^{T}\sum_{k}^{N_{ch}} EBP_t \cdot P_{k,t}^{ch} - \sum_{t=1}^{T}\sum_{k}^{N_{dis}} ESP_t \cdot P_{k,t}^{dis}
\end{aligned}
\tag{6}
$$

The first line of Label (6) represents the MG operator's revenue from selling energy to the consumers. The second line stands for the fuel cost of generation units and the start-up/shut-down costs. The third line denotes the costs associated with energy provided from the wind turbine (WT) and photovoltaic (PV) units. Here, it is assumed that the MG operator is not the owner of the renewable resources and is only responsible for the scheduling of the renewable units in the MG, so he should pay for energy provided by WT and PV. Finally, the last line expresses the costs associated with charge and discharge of EVs.

The second objective is to minimize the MG pollutants' emissions generated by DG units that consist of CO_2, NO_x and SO_2:

$$F_2 = \sum_{t=1}^{T} \left[\sum_{i=1}^{N_G} \left(Emi_{t,i}^{CO_2} + Emi_{t,i}^{NO_x} + Emi_{t,i}^{SO_2} \right) \right]. \tag{7}$$

The cost corresponding to MG frequency deviation should be minimized during scheduling horizon, which can be considered as a quadratic function of SMI as shown in Figure 4.

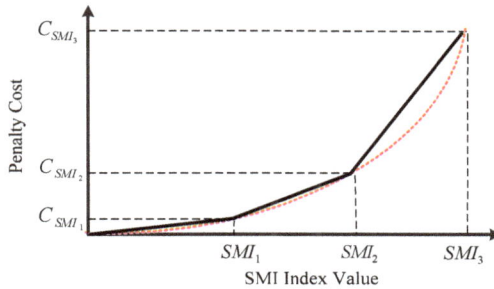

Figure 4. Piecewise linear cost curve of frequency deviation for an hour.

The piecewise linear model for one period is represented as the following:

$$f_3(t) = \sum_{m=1}^{NS(i)} v_{m,t} \cdot SMI_t \cdot u_t \quad \forall t, \tag{8}$$

$$F_3 = \sum_{t=1}^{T} f_3(t), \tag{9}$$

where $v_{m,t}$ is the slope of segment m in linearized total penalty cost and $NS(i)$ is the number of segments. In addition, u_t is a binary variable, equal to 0 if SMI has its expected limitation value (system stability is not at risk); otherwise, it is 1.

3.2. Mixed-Objective Function

Considering the above-mentioned objectives, the mixed-objective optimization problem can be developed as follows:

$$Max: \ Mobj = W_C(F_1) + W_E \cdot \mu_E \cdot (-F_2) + W_{SMI} \cdot (-F_3), \tag{10}$$

where W_C, W_E and W_{SMI} are weighting factors of the objective functions F_1, F_2 and F_3, respectively, and μ_E is the emission penalty factor in terms of £/kg. In the proposed weighted-sum model, the weighting factors can be set based on a multiple-criteria decision analysis (MCDA) done by the MG operator.

3.3. Constraints

The mentioned optimization problem is solved subject to the constraints as follows:

- Demand–supply balance equation: the balance between the total active power production and consumption in both grid-connected and isolated modes of MG is presented as:

$$P_t^{Grid} + \sum_{i=1}^{N_G} P_{i,t} + \sum_{w=1}^{N_W} P_{w,t} + \sum_{p=1}^{N_p} P_{p,t} + \sum_{k=1}^{N_{dis}} P_{k,t}^{dis} = \sum_{j=1}^{N_l} D_{j,t} + \sum_{k=1}^{N_{ch}} P_{k,t}^{ch}. \tag{11}$$

The left-side of Label (11) corresponds to the total available power in the MG at time period *t* including the expected WT and PV generation, the scheduled discharged power from EVs, the power of DG units and the exchanged power with the utility. In addition, the right side of Label (11) represents the total load and the power fed to the EVs.

- EVs constraints: Equations (12) and (13) define the power bounds for both EVs charging and discharging processes:

$$0 \leq P_{k,t}^{ch} \leq P_k^{\max,ch}, \tag{12}$$

$$0 \leq P_{k,t}^{dis} \leq P_k^{\max,dis}, \tag{13}$$

where $P_{k,t}^{ch}$ and $P_{k,t}^{dis}$ are charging and discharging power of *k*th EV. In addition, $P_k^{\max,ch}$ and $P_k^{\max,dis}$ are maximum charging and discharging power of *k*th EV, respectively.

- Power generation capacity: Active power output of a generation unit should be bounded within a range as follows:

$$P_{i,t} \leq P_i^{\max} u_{i,t} - R_{i,t}^{U} \quad \forall i, \forall t, \tag{14}$$

$$P_{i,t} \geq P_i^{\min} u_{i,t} + R_{i,t}^{D} \quad \forall i, \forall t \tag{15}$$

4. Simulation Results and Discussion

4.1. Case Study

The considered MG test system shown in Figure 5 is used to demonstrate the effectiveness of the proposed strategy. The test system includes a PV plant and a WT unit as renewable resources, a diesel engine as a DG, EVs and loads. Moreover, a dump load bank is considered for dumping the surplus energy produced by the RESs' units in isolated mode operation. Based on an economical assessment performed by HOMER Pro®, the optimal design of MG's sources (installation capacities of resources) is obtained for a region in the east of Iran (32.8649° N, 59.2262° E). With respect to this assessment and considering the technical constraints, the installed capacity of PV, WT and DG are calculated as 300 kW, 375 kVA and 300 kVA, respectively.

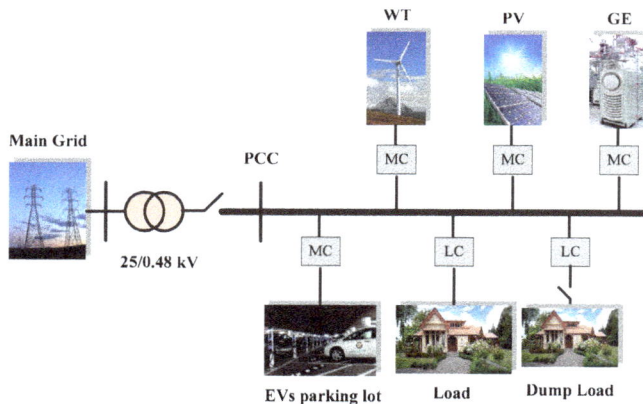

Figure 5. Single line diagram of the examined MG.

The forecasted load, output power of WT, PV and the summation power of WT and PV (P_{Ren}) in a typical day are depicted in Figure 6. It is also assumed that there are 100 EVs plugged into two different parking lots (in residential and office buildings) and their arrival times are modeled based on a Gaussian distribution with $\mu = 19$ and $\delta^2 = 10$ [21].

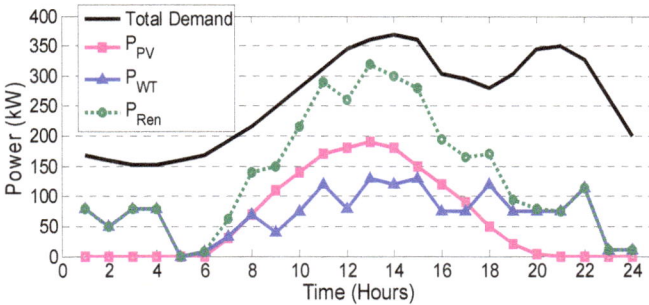

Figure 6. Forecasted demand and renewable generations in a typical day on an hourly basis.

Here, the minimum and maximum limit of *SOC* are considered as 40% and 90% of the total battery capacity, respectively [22]. The connected EVs are assumed to be charged at different current ratings (i.e., 2, 10 or 30 Ampere), and to be discharged either with 2 or 10 Ampere [23].

Moreover, it is assumed that the charge/discharge processes in both parking lots is the same and the number of EVs in each parking lot is evaluated based on the available profiles depicted in Figure 7. Due to EVs travelling, the number of EVs in both parking lots is a percentage of the total number of EVs.

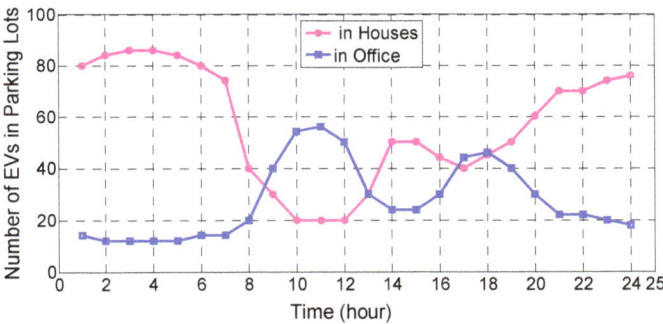

Figure 7. Number of EVs in the house and office parking lots.

4.2. EVs Charge/Discharge Process with/without EVPC

Based on the explanation in Section 2, the DNS_{Ren} and SMI index over the 24-h horizon are achieved in the under study system that are shown in Figures 8 and 9, respectively. To obtain the cost signal based on the SMI index, the minimum and maximum values of SMI correspond with the minimum and maximum values of cost signal in the wholesale market with the application of numerical analysis. The other values of cost signal should remain in this range in such a way to follow the SMI index. It should be noted that this cost signal is applied in the isolated mode by the MG operator, but in the connected mode, the MG operator is a price taker and the cost signal of the wholesale market is used for EV management.

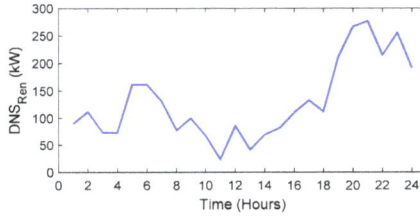

Figure 8. The demand not supplied by renewable resources (DNS_{Ren}) of MG over the 24-h horizon.

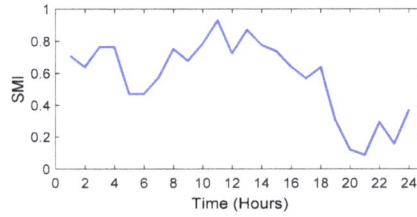

Figure 9. SMI over the 24-h horizon.

In this case, as shown in Figure 10a, the minimum electricity buying and selling prices are 0.186 and 0.184 (£/kWh), and maximum electricity buying and selling prices are 0.202 and 0.212 (£/kWh), respectively. Figure 10b shows the day-ahead electricity selling and buying prices as adopted from [24]. As can be seen, the minimum electricity buying and selling prices are set to 0.169 and 0.149 (£/kWh), while the maximum electricity buying and selling prices are considered as 0.191 and 0.245 (£/kWh), respectively. This cost signal is used to be compared with the numerical results of the proposed signal. The exchanged power between EVs and the MG with/without EVPC in each time interval is shown in Figure 11 for a typical day.

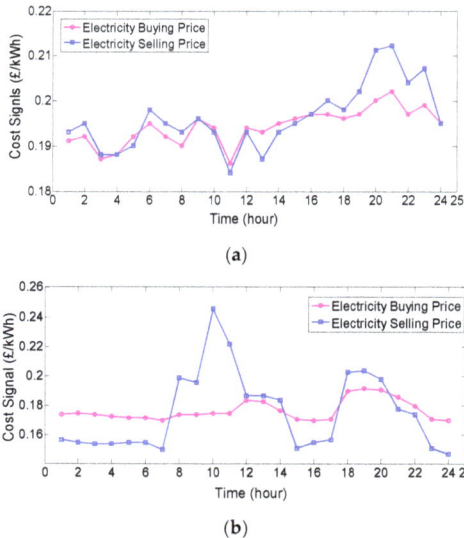

(**a**)

(**b**)

Figure 10. Electricity buying/selling price (**a**) based on power mismatch between load and P_{Ren} (**b**) extracted from [23].

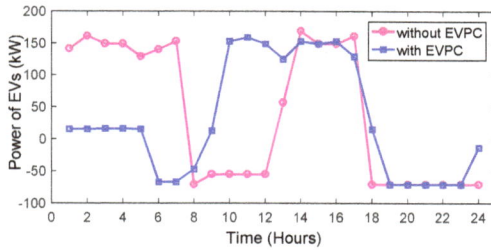

Figure 11. The exchanged power between EVs and the MG with/without EVPC.

4.3. Frequency Stability Analysis with/without EVPC

To investigate the effect of EVPC on the MG frequency stability, it is assumed that at time 1:00, the MG switches into isolated mode and remains in this mode during the scheduling horizon. Due to a fault occurrence in the upstream at time 1:00, the MG switches into the isolated mode. In isolated mode, MG frequency response is studied in two cases. In the first case, EVs do not participate in load frequency control, while, in the second one, EVs contribute in the MG frequency regulation. Figure 12 shows the MG frequency variations in two cases.

Comparison of the two cases shows that the application of EVPC results in a lower frequency deviation. As can be seen in the same figure, with EVPC, frequency varies in the range between 59 and 61.4 Hz; however, without it, the frequency drops to less than 57 Hz when the MG enters the isolated mode. As can be observed from Figure 13, the minimum value of SMI occurs at 6:00. It can be shown that, at this time, the system is in the critical condition without the application of EVPC.

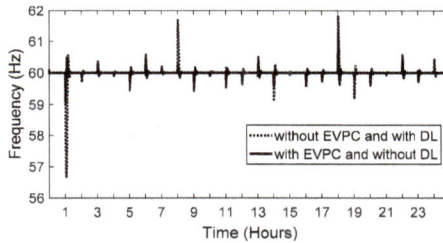

Figure 12. Frequency variation in two cases over a 24-h horizon.

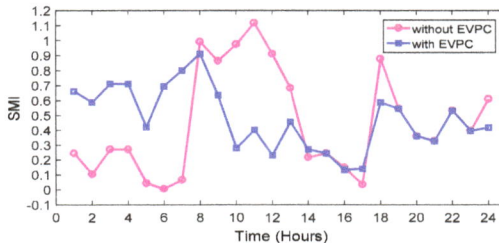

Figure 13. Variation of SMI with and without EVPC.

During islanded operation, due to low inertia of MG and intermittent nature of RESs and load, there might be some frequency deviations beyond the acceptable range. As depicted in Figure 14, due to a sudden load change of 20 kW at 6:30, the system frequency becomes unstable without EVPC.

At this time, EVs are charged 140 kW without EVPC; however, with the application of that, they are discharged at the rate of 68 kW/h. In fact, when the load demand increases suddenly, the amount of reserve reduces. In this condition, generation of RESs is very low (7 kW) and so EVs' high charging leads to frequency drop. However, with the application of EVPC, due to a low amount of reserve and the SMI index, EVs assist in the discharge process and the system obtains a proper margin for its reserve and, consequently, frequency remains within its limitation. It should be mentioned that, without EVPC, the frequency instability occurs at 11:00 due to surplus energy produced by RESs.

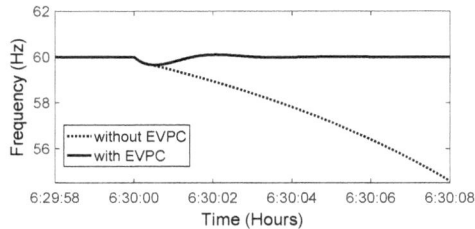

Figure 14. Frequency variation at 6:00 (SMI is in its lowest value).

In order to assess fluctuation of frequency accurately, the frequency deviation with and without EVPC is shown in Figure 15. At this time, without the EVPC scheme, EVs are discharged up to 56 kW and with EVPC EVs are charged 142 kW, accordingly. On the other hand, with EVPC, EVs are charged with lower power and frequency variation is less than the other case where there is no EVPC.

Thus, the value of SMI without EVPC is more than 1 and the system is unstable at 11:00. However, with the application of DL, frequency remains within its expected value. In this case, DL consists of a bank of resistive loads each with a consumption level of 1.75 kW (up to 175 kW). As it can be observed from Figure 16, to keep the system power balance at 11:00, 30 kW of DL activated. Thus, with the application of DL in the case without EVPC, the system remains stable at all times.

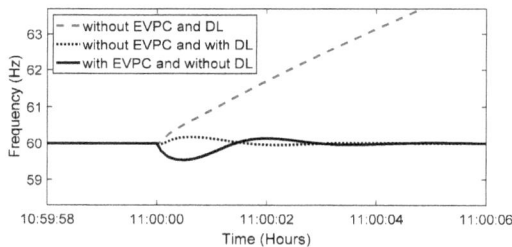

Figure 15. Frequency variation.

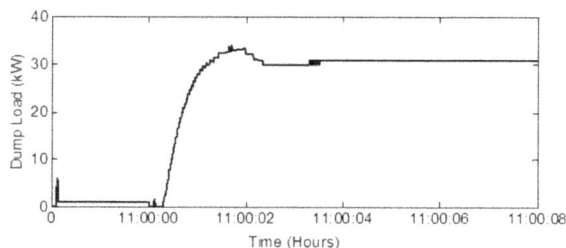

Figure 16. Dump load activation at 11:00.

The maximum load demand occurs at 14:00 (368 kW), and frequency variation at this time is shown in Figure 17. At this time, load has its maximum value and the production of RESs is 300 kW. Based on both cost signals, EVs are charged 152 kW and 168 kW in cases with and without EVPC, respectively, and the frequency is kept in its limitations in both cases.

As can be observed in Figure 12, frequency has an overshoot at 8:00 and 18:00 that is zoomed in on in Figure 18. At these times, since SMI is higher in conditions without EVPC than that of in circumstances with EVPC; hence, frequency deviation is further in the former conditions than the latter ones.

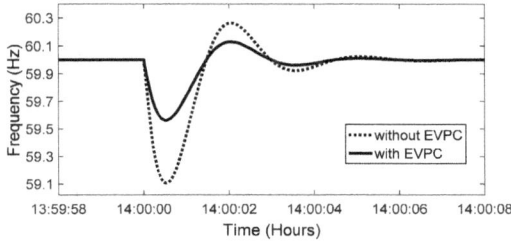

Figure 17. Frequency variation at the maximum load.

The optimization results of the under-study system with and without the application of EVPC are presented in Table 1. As it can be seen, the implementation of EVPC achieves better economic results as compared to the case without it. The results reveal that the MG operator's profit (F_1) is 324.2 £, which is 30.1 £ more than the case without EVPC. Moreover, MG emission is decreased about 205.849 kg with the proposed strategy.

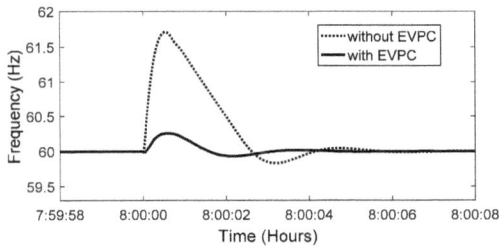

(a)

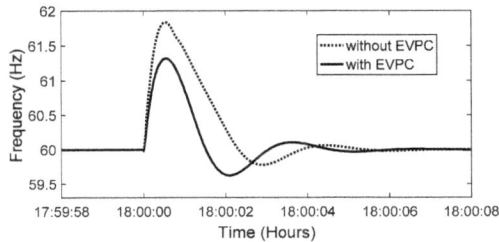

(b)

Figure 18. Frequency variation (**a**) at 8:00 and (**b**) at 18:00.

Furthermore, since the frequency deviations in the case with EVPC remain in the allowable range, the cost corresponding with load frequency control is zero. However, in the case without EVPC, cost of frequency control (F_3) raises up to 11.7 £. Finally, the objective function that is equal to the total MG operator's profits with and without the implementation of EVPC are obtained as 282.1 and 237.6 £, respectively. Thus, the assessment of the results shows that, with the application of EVPC, the total MG operator's profit is increased considerably.

Table 1. The optimization results of microgrid (MG) in scheduling time horizon.

Case	F_1 (£)	F_2 (kg)	F_3 (£)	Objective Function (£)
Without EVPC	293.6	3408.691	11.7	237.6
With EVPC	324.2	3202.842	0.0	282.1

EVPC: Electric vehicle power controller.

5. Conclusions

In this paper, an optimal management strategy was proposed in order to schedule the EVs' charging/discharging process with the goal of improving frequency stability of MG during autonomous operating conditions. In this way, a cost signal including EBP and ESP was proposed based on the variation of SMI index that followed the intermittent nature of RESs and load variation. Based on the proposed strategy, when frequency is at risk, EVs could absorb (inject) the surplus (shortage) of energy and act as energy storage systems. The results also showed, that with the application of this strategy, the frequency variation of MG in isolated mode is less than that without it. Moreover, it is not required to use DL when the EVPC strategy is applied in the MG. Furthermore, it was understood from the results that using EVPC strategy could increase the total MG operator's profit and decrease the emission substantially.

Author Contributions: Mostafa Vahedipour-Dahraie and Homa Rashidizaheh-Kermani developed the model; Mostafa Vahedipour-Dahraie simulated the case studies; Mostafa Vahedipour-Dahraie, Homa Rashidizaheh-Kermani and Amjad Anvari-Moghaddam analyzed the data; Mostafa Vahedipour-Dahraie, Homa Rashidizaheh-Kermani and Hamid Reza Najafi wrote the manuscript; Amjad Anvari-Moghaddam and Hamid Reza Najafi and Josep M. Guerrero provided their comments on the paper.

Conflicts of Interest: The authors declare no conflict of interest.

Nomenclature

N_G	Number of generation units
N_J	Set of loads number
N_s	Number of scenarios
N_W	Number of wind turbine units
N_P	Number of photovoltaic units
N_K	Number of EVs
T	Scheduling time (24 h a day)
$i(j)$	Index of generating units (loads), running from 1 to N_G (N_J)
b,n,r	Indices of system buses
t (s)	Index of time (scenario), running from 1 to T (N_s)
w (p)	Index of WT (PV) units, running from 1 to N_W (N_P)
k	Index of EVs, running from 1 to N_K
$C_{w,t}$ ($C_{p,t}$)	Energy bid submitted by WT w (PV p) in period t (£/kWh)
$C_{i,t}^{RU}$ ($C_{i,t}^{RD}$)	Bid of the up (down)-spinning reserve submitted by unit i in period t (£/kWh)
$C_{j,t}^{RU}$ ($C_{j,t}^{RD}$)	Bid of the up (down)-spinning reserve submitted by load j in period t (£/kWh)
$C_{i,t}^{RNS}$	Bid of the non-spinning reserve submitted by unit i in period t (£/kWh)
$SUC_{j,t}$ ($SDC_{j,t}$)	Start-up (shut-down) cost of unit i in period t (£)
$\rho(t)$	Electricity price in period t (£)

$P_{i,t}$	Scheduled power for unit i in period t (kW)
$P_{w,t}$ $(P_{v,t})$	Output power of WT w (PV v) in period t (kW)
$P_x{}^{max}$ $(P_x{}^{min})$	Maximum (minimum) generating capacity of unit x (kW)
$P_{k,t}{}^{EV,ch}$ $(P_{k,t}{}^{EV,dis})$	Charging (discharging) and discharging power of EV k in period i (kW)
$R_{i,t}^U$ $(R_{i,t}^D)$	Scheduled up (down)-spinning reserve for unit i in period t (kW)
$R_{i,t}^{NS}$	Scheduled non-spinning reserve for unit i in period t (kW)
$R_{j,t}^U$ $(R_{j,t}^D)$	Scheduled up (down)-spinning reserve for load j in period t (kW)
$u_{i,t}$	Binary variable, equal to 1 if unit i is scheduled to be committed in period t otherwise 0
$y_{i,t}$	Binary variable, equal to 1 if unit i is starting up in period t otherwise 0

References

1. Gholami, A.; Aminifar, F.; Shahidehpour, M. Front lines against the darkness: Enhancing the resilience of the electricity grid through microgrid facilities. *IEEE Electrific. Mag.* **2016**, *4*, 18–24. [CrossRef]
2. Khodaei, A. Provisional microgrids. *IEEE Trans. Smart Grid* **2015**, *6*, 1107–1115. [CrossRef]
3. Guerrero, J.M.; Vasquez, J.C.; Matas, J.; de Vicuna, L.G.; Castilla, M. Hierarchical control of droop-controlled AC and DC microgrids-A general approach towards standardization. *IEEE Trans. Ind. Electron.* **2011**, *58*, 158–172. [CrossRef]
4. Pascal, M.; Rachid, C.; Alexandre, O. Optimizing a battery energy storage system for frequency control application in an isolated power system. *IEEE Trans. Power Syst.* **2009**, *24*, 1469–1477.
5. Mu, Y.; Wu, J.; Ekanayake, J.; Jenkins, N.; Jia, H. Primary Frequency Response from Electric Vehicles in the Great Britain Power System. *IEEE Trans. Smart Grid* **2013**, *4*, 1142–1150. [CrossRef]
6. Rotering, N.; Ilic, M. Optimal charge control of plug-in hybrid electric vehicles in deregulated electricity markets. *IEEE Trans. Power Syst.* **2010**, *26*, 1021–1029. [CrossRef]
7. Vahedipour-Dahraie, M.; Najafi, H.R.; Anvari-Moghaddam, A.; Guerrero, J.M. Study of the Effect of Time-Based Rate Demand Response Programs on Stochastic Day-Ahead Energy and Reserve Scheduling in Islanded Residential Microgrids. *Appl. Sci.* **2017**, *7*, 378. [CrossRef]
8. Guo, F.; Inoa, E.; Choi, W.; Wang, J. Study on global optimization and control strategy development for a PHEV charging facility. *IEEE Trans. Veh. Technol.* **2012**, *61*, 2431–2441. [CrossRef]
9. Chukwu, U.C.; Mahajan, S.M. Real-Time Management of Power Systems with V2G Facility for Smart-Grid Applications. *IEEE Trans. Sustain. Energy* **2013**, *4*, 1142–1150. [CrossRef]
10. Nienhueser, I.A.; Qiu, Y. Economic and environmental impacts of providing renewable energy for electric vehicle charging—A choice experiment study. *Appl. Energy* **2106**, *18*, 256–268. [CrossRef]
11. Gao, S.; Chau, K.T.; Liu, C.; Wu, D.; Chan, C.C. Integrated Energy Management of Plug-in Electric Vehicles in Power Grid with Renewables. *IEEE Trans. Veh. Tecnol.* **2014**, *63*, 3019–3027. [CrossRef]
12. Zhang, M.; Chen, J. The Energy Management and Optimized Operation of Electric Vehicles Based on Microgrid. *IEEE Trans. Power Deliv.* **2014**, *29*, 1427–1435. [CrossRef]
13. Wu, T.; Yang, Q.; Bao, Z.; Yan, W. Coordinated energy dispatching in microgrid with wind power generation and plug-in electric vehicles. *IEEE Trans. Smart Grid* **2013**, *4*, 1453–1463. [CrossRef]
14. Ghofrani, M.; Arabali, A.; Etezadi-Amoli, M.; Fadali, M.S. Smart Scheduling and Cost-Benefit Analysis of Grid-Enabled Electric Vehicles for Wind Power Integration. *IEEE Trans. Smart Grid* **2014**, *5*, 2306–2313. [CrossRef]
15. Izadkhast, S.; Garcia-Gonzalez, P.; Frias, P. An aggregate model of plug-in electric vehicles for primary frequency control. *IEEE Trans. Power Syst.* **2015**, *30*, 1475–1482. [CrossRef]
16. Almeida, P.M.R.; Lopes, J.A.P.; Soares, F.J.; Seca, L. Electric vehicles participating in frequency control: Operating islanded systems with large penetration of renewable power sources. In Proceedings of the 2011 IEEE Trondheim PowerTech, Trondheim, Norway, 19–23 June 2011; pp. 5–10.
17. O'Connell, N.; Wu, Q.; Stergaard, J.; Nielsen, A.H.; Cha, S.T.; Ding, Y. Day-ahead tariffs for the alleviation of distribution grid congestion from electric vehicles. *Electr. Power Syst. Res.* **2012**, *92*, 106–114.
18. Wu, C.; Mohsenian-Rad, H.; Huang, J. Vehicle-to-Aggregator Interaction Game. *IEEE Trans. Smart Grid* **2012**, *3*, 434–442. [CrossRef]
19. Galus, M.D.; Koch Andersson, S. Provision of load frequency control by PHEV, controllable loads, and a cogeneration unit. *IEEE Trans. Ind. Electron.* **2011**, *58*, 4568–4582. [CrossRef]

20. Ma, Y.; Houghton, T.; Cruden, A.; Infield, D. Modeling the Benefits of Vehicle-to-Grid Technology to a Power System. *IEEE Trans Power Syst.* **2012**, *27*, 1012–1020. [CrossRef]
21. Rassaei, F.; Soh, W.S.; Chua, K.C. Demand Response for Residential Electric Vehicles with Random Usage Patterns in Smart Grids. *IEEE Trans. Sustain. Energy* **2015**, *6*, 1367–1376. [CrossRef]
22. Pillai, J.R.; Bak-Jensen, B. Integration of vehicle-to-grid in the Western Danish power system. *IEEE Trans. Sustain. Energy* **2011**, *2*, 12–19. [CrossRef]
23. Vahedipour-Dahraie, M.; Rashidizadeh-Kermani, H.; Najafi, H.R. A Proposed Strategy to Manage Charge/Discharge of EVs in a Microgrid Including Renewable Resources. In Proceedings of the 24th Iranian Conference on Electrical Engineering (ICEE), Shiraz, Iran, 10–12 May 2016; pp. 649–654.
24. UK National Grid. Available online: http://www.bmreports.com/ (accessed on 10 May 2016).

applied
sciences

MDPI

Article

Study of the Effect of Time-Based Rate Demand Response Programs on Stochastic Day-Ahead Energy and Reserve Scheduling in Islanded Residential Microgrids

Mostafa Vahedipour-Dahraie [1], Hamid Reza Najafi [1,*], Amjad Anvari-Moghaddam [2] and Josep M. Guerrero [2]

[1] Department of Electrical & Computer Engineering, University of Birjand, Birjand 9856, Iran; vahedipour_m@birjand.ac.ir
[2] Department of Energy Technology, Aalborg University, Aalborg East 9220, Denmark; aam@et.aau.dk (A.A.-M.); joz@et.aau.dk (J.M.G.)
*   Correspondence: h.r.najafi@birjand.ac.ir; Tel.: +98-56-3220-2049

Academic Editor: Antonio Ficarella
Received: 8 February 2017; Accepted: 7 April 2017; Published: 11 April 2017

Abstract: In recent deregulated power systems, demand response (DR) has become one of the most cost-effective and efficient solutions for smoothing the load profile when the system is under stress. By participating in DR programs, customers are able to change their energy consumption habits in response to energy price changes and get incentives in return. In this paper, we study the effect of various time-based rate (TBR) programs on the stochastic day-ahead energy and reserve scheduling in residential islanded microgrids (MGs). An effective approach is presented to schedule both energy and reserve in presence of renewable energy resources (RESs) and electric vehicles (EVs). An economic model of responsive load is also proposed on the basis of elasticity factor to model the behavior of customers participating in various DR programs. A two-stage stochastic programming model is developed accordingly to minimize the expected cost of MG under different TBR programs. To verify the effectiveness and applicability of the proposed approach, a number of simulations are performed under different scenarios using real data; and the impact of TBR-DR actions on energy and reserve scheduling are studied and compared subsequently.

Keywords: demand response (DR); scheduling; time-based rate (TBR) programs; renewable energy resources (RESs); electric vehicles (EVs)

1. Introduction

One of the major thrust areas of demand side management (DSM) is demand response (DR) which is defined as a set of actions taken to reduce users' electricity consumptions in response to higher market prices or market incentives [1]. Moreover, system operators may apply DR programs to reduce the load temporarily in emergency grid conditions such as unit outage or unpredictable change in renewable generation [2,3]. Therefore, the main idea of DR is to encourage customers to manage their consumption patterns in a way not only to maximize their own utility, but also to support safe operation of the power system [4].

According to the Federal Energy Regulatory Commission (FERC), DR programs can be classified into two major categories, namely, time-based rate (TBR) and incentive-based programs (IBPs) [5]. In TBR programs (also known as price-based DR programs [6]), time-varying prices are given to consumers based on the electricity price in different time periods, encouraging them to change their consumption level in response to the changing price signals. On the other hand, in IBP schemes, customers are offered fixed or time-varying incentives, to reduce their electricity consumption during periods of system stress, however they would be penalized for no participation in the program [7].

The focus of this paper is on TBR programs which are mainly divided into three categories, real-time pricing (RTP), time of use (TOU), and critical peak pricing (CPP) programs. These programs are well-suited for implementation in residential areas (e.g., residential microgrids (MGs)) where there are more possibilities for load management purposes [8–11]. However, there exist a number of challenges, such as rebound peaks during low cost periods and service interruptions. Moreover, the presence of uncertain elements within an environment such as wind and photovoltaic (PV) power generation imposes development of sophisticated balancing mechanisms between supply and demand to meet the system stability. Therefore, TBR models need to be well-designed and implemented to provide efficient operating conditions for such systems in presence of uncertainties.

Regarding the MG scheduling under uncertainty, much research has been done recently [12–21]. The effect of wind energy forecast errors on the network-constrained market-clearing problem, in which energy and reserve are simultaneously dispatched, was investigated in [12,13] using two-stage stochastic programming models. Based on the proposed method in [12], cost of MG was minimized with regard to the uncertainty of renewable energy resources (RESs). In [13], optimal dispatch of a MG was presented with regard to emissions and fuel consumption cost minimization using heuristic optimization. Authors in [14,15] exploited MG management as a multi-objective optimization problem to mitigate emission level as well as operation and maintenance costs. To obtain efficient energy management, artificial intelligence techniques were also used with multi-objective optimization programming [16]. However, in the reviewed literature, the procurement of the MG reserve (in terms of spinning and/or non-spinning reserve) for reliable operation of the system has been neglected. To address this issue, effective methods for providing reserve in typical MGs with high penetration of RESs are developed based on DR programs [17–19]. In [17], a day-ahead market structure was presented where DR can provide contingency reserves through a bidding procedure representing the cost of load curtailment. Also, authors in [18] introduced a price-responsive DR action for optimal regulation service reserve provision under high levels of wind penetration. The same type of study was carried out in [19], considering load uncertainty and generation unavailability as different working scenarios. In view of the problem-solving strategies, most of the reviewed research works have utilized stochastic programming techniques, however some have applied other methods such as robust optimization or Monte Carlo simulation [20,21]. In [22], a stochastic AC security-constrained unit commitment problem under wind power uncertainty has been formulated. Also, a stochastic multi-objective framework has been proposed in [23], for joint energy and reserve scheduling in day-ahead however, this reference has not considered AC network, load, EVs and wind power uncertainties. Furthermore, authors in [24] have proposed a multi-objective structure that can optimize objective functions including operation costs of MG, but they have not considered demand and EVs uncertainty in day-ahead scheduling.

This paper presents the effect of different types of TBR programs on the MG operation costs and shaping the load profile in presence of RESs and electrical vehicles (EVs). EVs are employed for energy scheduling or peak shaving with fast charging and discharging capabilities, while the responsive loads are used to supply a part of the required MG reserve to compensate RESs uncertainties. Monte-Carlo simulations together with k-means clustering technique are applied to create several scenarios corresponding to renewable generation variations and EVs owners' behaviors. The generated scenarios are then reduced and fed into a two-stage optimization model developed for minimizing the operation costs. In the first stage of optimization, the energy and reserve costs are minimized simultaneously and in the second stage, the cost associated with the rescheduling of generating units (due to the variations in wind turbine (WT) and photovoltaic (PV) output powers) is minimized. Finally, simulation results for co-optimization of energy and reserves in the examined residential MG are presented and compared under different DR programs and operating conditions. As a whole, the main contributions of this paper can be highlighted as:

- Optimal management of an islanded MG with RTP-based DR programs using a scenario-based two-stage stochastic programming model.
- Simultaneous energy and reserve scheduling of MGs with regard to different DR schemes in an uncertain environment.
- Assessment of TBR-based DR programs under different scenarios with/without considering EVs participation.

The remainder of this paper is arranged as follows: a network-constrained day-ahead market clearing model is introduced in Section 2 and it is reformulated into a mixed integer programming (MIP) model in Section 3. The case studies are presented in Section 4 and the simulation results are discussed thereafter. Finally, Section 5 concludes this paper with future scope.

2. Model Description

A network-constrained day-ahead market clearing model is developed under a two-stage stochastic programming framework in order to accommodate the uncertain nature of RESs and EVs. Based on [25] the MG uncertainties can be categorized into two groups:

(1) Normal operation uncertainties (including errors in forecasting wind data, EV operation, and real-time market prices).
(2) Contingency-based uncertainties (including random forced outages, unintentional islanding, and resynchronization events).

The subject area of this paper mainly falls in the first category so the optimization model is developed in a way to effectively consider normal operation uncertainties including forecasting errors of WT and PV power production and EV owner behaviors. A set of scenarios representing MG uncertainties are generated for scheduling horizon. In order to render the problem tractable, an appropriate scenario-reduction algorithm is applied to reduce the generated scenarios into an optimal subset that represents well enough the uncertainties. In the next step, the optimization problem is solved in two stages using commercially-available software packages. In the first stage of the proposed optimization model, energy and reserves are jointly scheduled to balance supply and demand. The second stage corresponds to operation management in several actual MG modes and deals with variables that are scenario-dependent and have different values for every single scenario. In other words, the first stage corresponds to the optimal decision for the deterministic base case, while the second stage examines the feasibility and optimality of the first stage decisions under system contingencies.

In the proposed framework, different customers sign contracts for participating in various TBR programs and submit them to the MG operator. Based on the type of the consumers' contributions, MG operator finds the optimal day-ahead energy and reserve scheduling with regard to the minimum expected cost of operation. At the same time, optimum participation level of consumers in each DR program for reserve procurement is determined. Also, MG operator schedules the charging and discharging process of the EVs for any time intervals in the studied period.

2.1. Market-Based DR Model

In order to evaluate the impact of residential customers' participation in DR programs on load profile characteristics, an economic model of responsive loads is developed on the basis of elasticity factors. Elasticity is defined as demand sensitivity with respect to the electricity prices [26].

$$E(t,t) = \frac{\rho_0(t)}{D(t)} \frac{\partial D(t)}{\partial \rho(t)} \tag{1}$$

where, $D(t)$ and $\rho_0(t)$ are the nominal/initial value of demand and electricity price, respectively. Based on a single-period elastic load model, the customer changes his demand to achieve the maximum benefit from $D(t)$ to $D_{DR}(t)$ as:

$$D_{DR}(t) = D(t) + \Delta D(t) \tag{2}$$

The customer benefit for the tth time interval can be calculated as:

$$S(D_{DR}(t)) = B(D_{DR}(t)) - D_{DR}(t) \cdot \rho(t) \tag{3}$$

where, $S(D_{DR}(t))$ and $B(D_{DR}(t))$ represent customer benefit and income at time t after implementing DR programs, respectively. In order to maximize customer benefit, the following condition must be met [27]:

$$\frac{\partial S(D_{DR}(t))}{\partial D_{DR}(t)} = 0 \Rightarrow \frac{\partial B(D_{DR}(t))}{\partial D_{DR}(t)} = \rho(t) \tag{4}$$

Therefore, the customer utility function would get a quadratic form as follows [27]:

$$B(D_{DR}(t)) = B_0(t) + \rho_0(t)[D_{DR}(t) - D(t)] \times \left[1 + \frac{D_{DR}(t) - D(t)}{2E(t,t) \cdot D(t)}\right] \tag{5}$$

Differentiating (5) with respect to $D_{DR}(t)$ and substituting the result in (4) yields:

$$\rho(t) = \rho_0(t) \cdot \left[1 + \frac{D_{DR}(t) - D(t)}{E(t,t) \cdot D(t)}\right] \tag{6}$$

Therefore, a customer's consumption behavior over the time can be obtained as follows:

$$D_{DR}(t) = D(t) \cdot \left[1 + E(t,t) \cdot \frac{\rho(t) - \rho_0(t)}{\rho_0(t)}\right] \tag{7}$$

In a multi-period elastic loads model, the price elasticity of the tth period versus the hth period can be defined as [26]:

$$E(t,h) = \frac{\rho_0(h)}{D_0(t)} \cdot \frac{\partial D(t)}{\partial \rho(h)} \tag{8}$$

Considering the linear relationship between the hourly demand level and the electricity prices, it can be expressed that:

$$D_{DR}(t) = D(t) \cdot \left[1 + \sum_{\substack{t=1 \\ t \neq h}}^{T} E(t,h) \cdot \frac{\rho(h) - \rho_0(h)}{\rho_0(h)}\right] \tag{9}$$

Combining (7) and (9), the responsive load economic model can be extracted as follows:

$$D_{DR}(t) = D(t) \cdot \left[1 + E(t,t) \cdot \frac{\rho(t) - \rho_0(t)}{\rho_0(t)} + \sum_{\substack{h=1 \\ h \neq t}}^{T} E(t,h) \cdot \frac{\rho(h) - \rho_0(h)}{\rho_0(h)}\right] \tag{10}$$

2.2. EVs Participation in DR Programs

EVs can be considered in three different modes: grid-connected mode, idle mode, or driving mode. In grid-connected mode, the MG operator can schedule charging/discharging process of EVs batteries. EVs can exchange power with the MG based on their state of charge (SOC), stop time in the parking lot (PL) and the electricity price in each DR program. In this case, EVs are considered to be probabilistic loads or generations which can be evaluated by stochastic methods [28]. The exchange power between each EV and the network can be obtained as [28]:

$$P_{k,t}^{EV} = \eta_c P_{k,t}^c - \frac{P_{k,t}^d}{\eta_d} \quad \forall t \in u_k \tag{11}$$

The SOC of EVs connected to the network is updated by Equation (12) [28].

$$BC_k \cdot SOC_{k,t} = BC_k \cdot SOC_{k,t-1} + P_{k,t-1}^{EV} \quad \forall t \in T, \forall k \in Nk \tag{12}$$

where $SOC_{k,t-1} = SOC_{k,I}$, if $t = 1$. BC_k is battery capacity of EV in kWh and $SOC_{k,I}$ is the initial SOC of kth EV. It is important to control the charge and discharge energy of the parked vehicle w such that the *SOC* of the battery could be kept within the allowed range SOC_k^{min} and SOC_k^{max}.

Besides, in idle or driving mode, there is no power exchange between EV and the network, however the stored energy might decrease depending on the EV trip length (L_k) and its energy consumption rate (r_k). It is assumed that each EV returns to the PL after driving L_k km and is plugged back into the network. Thus, the SOC at the time of arrival (SOC_k^{ent}) can be estimated by Equation (13) [28].

$$SOC_k^{ent} = SOC_k^{int} - L_k \times r_k \quad \forall k \in Nk \tag{13}$$

where, SOC_k^{int} is the initial SOC at the beginning of the trip.

2.3. Renewable Energy Resources

Output power of WT and PV plants are inherently intermittent. In order to model the stochastic wind speed (and the WT behavior accordingly), the divided Weibull probability density function (PDF) is usually employed. The general Weibull PDF of wind speed can be formulated as follows [29]:

$$PDF(v) = \frac{k}{c} \left(\frac{v}{c}\right)^{k-1} \cdot e^{-\left(\frac{v}{c}\right)^k} \tag{14}$$

where v, k and c are wind speed, shape factor (dimensionless) and scale factor, respectively.

Besides, the output power of WT can be described by Equation (15) [30]:

$$P_w(v) = P_w^r \cdot \begin{cases} 0 & ; 0 \leq v \leq v_{in} \ and \ v \geq v_{out} \\ \frac{v_{in}^3}{v_{in}^3 - v_r^3} + \frac{bv^3}{v_r^3 - v_{in}^3} & ; v_{in} \leq v \leq v_r \\ 1 & ; v_r \leq v \leq v_{out} \end{cases} \tag{15}$$

where v_r, v_{in} and v_{out} indicate the rated speed, cut-in speed and cut-out speed of the WT, respectively, and P_w^r represents the total rated power of WT.

The distribution of hourly irradiance usually follows a bimodal distribution, which can be seen as a linear combination of two unimodal distribution functions [31,32]. A Beta PDF is utilized for each unimodal, as stated in the following [31]:

$$f_b(\varphi) = \begin{cases} \frac{\Gamma(\alpha+\beta)}{\Gamma(\alpha) \cdot \Gamma(\beta)} \cdot \varphi^{(\alpha-1)} \times (1-\varphi)^{\beta-1} & for \quad 0 \leq \varphi \leq 1, \ \alpha \geq 0, \ \beta \geq 0 \\ 0 & otherwise \end{cases} \tag{16}$$

The parameters of the Beta distribution function (α, β) are calculated based on the mean (μ) and standard deviation (σ) of the random variable [31].

In this paper, to model the uncertainties of output power for WT and PV units, a set of possible scenarios is generated based on Metropolis–Hastings algorithm and reduced thereafter to a number of distinct scenarios using the k-means clustering technique [33].

3. Optimization Problem Formulation

3.1. Objective Function

The objective function is defined based on the minimization of the total expected cost (EC) of an isolated residential MG which includes cost of energy and reserve provision as well as the operating cost in different working scenarios.

$$
\begin{aligned}
EC \quad &= \sum_{t=1}^{T} \sum_{i=1}^{Ng} \left[(A_i \cdot u_{i,t} + B_i \cdot P_{i,t}) + SUC_i \cdot y_{i,t} + SDC_i \cdot z_{i,t} \right. \\
&\left. + (C_{i,t}^{RD} \cdot R_{i,t}^{D} + C_{i,t}^{RU} \cdot R_{i,t}^{U} + C_{i,t}^{RNS} \cdot R_{i,t}^{NS}) \right] \\
&+ \sum_{w=1}^{Nw} C_{w,t} \cdot P_{w,t} + \sum_{p=1}^{Np} C_{p,t} \cdot P_{p,t} \\
&+ \sum_{k=1}^{Nk_d} SPR_t \cdot P_{k,t}^{d} - \sum_{k=1}^{Nk_c} BPR_t \cdot P_{k,t}^{c} \\
&+ \sum_{t=1}^{T} \sum_{j=1}^{Nj} C_{j,t}^{RD} \cdot R_{j,t}^{D} + C_{j,t}^{RU} \cdot R_{j,t}^{U} - \sum_{j=1}^{Nj} C_{j,t} \cdot L_{j,t} \\
&+ \sum_{s=1}^{Ns} \sum_{t=1}^{T} \sum_{i=1}^{Ng} \left[SUC_i \cdot (y_{i,t,s} - y_{i,t}) + SDC_i \cdot (z_{i,t,s} - z_{i,t}) \right. \\
&\left. + C_{i,t} \cdot (r_{i,t,s}^{U} + r_{i,t,s}^{NS} - r_{i,t,s}^{D}) \right] \\
&+ \sum_{s=1}^{Ns} \sum_{t=1}^{T} \left[\sum_{j=1}^{Nj} C_{j,t} \cdot (r_{j,t,s}^{U} - r_{j,t,s}^{D}) \right. \\
&\left. + \sum_{w=1}^{Nw} C_{w,t} \cdot \Delta P_{w,t,s} + \sum_{p}^{Np} C_{p,t} \cdot \Delta P_{p,t,s} \right] \\
&+ \sum_{j=1}^{Nj} V^{LOL} \cdot L_{j,t,s}^{shed}
\end{aligned}
\tag{17}
$$

In Equation (17), the first line of the objective function states the costs associated with energy provided from the generating units and the start-up and shut-down costs, and the second line expresses the commitment of the generating units to provide reserves. The third line denotes the costs associated with energy provided from the WT and PV units. The fourth line expresses the cost associated with charge/discharge of EVs and the fifth line considers the utility of the demand loads and their up and down reserve provision.

The rest of the terms in the objective function deal with the operating cost in different working scenarios. In this regard, the sixth and the seventh lines consider cost of unit commitment and the cost of deploying reserves from those units in different scenarios. The eighth line represents the cost of deploying reserves from DR programs and the ninth line states the costs associated with energy provided from WT and PV units. Here, it is assumed that the MG operator would pay for energy provided by WT and PV. Finally, the last term stands for the expected cost of energy not served for the inelastic loads.

3.2. Constraints

The problem constraints include two parts; the first-stage constraints and second-stage constraints. The first-stage ones are associated with the base case scenario (i.e., deterministic operating condition), and can be expressed as follows:

- Power balance in steady state; Equation (18) represents the active power balance in MG in steady state [21].

$$
\sum_{i=1}^{Ngb} P_{i,t} + \sum_{w=1}^{Nwb} P_{w,t} + \sum_{p=1}^{Np} P_{p,t} + \sum_{k=1}^{Nkb_d} P_{k,t}^{d} = L_{b,t} + \sum_{k=1}^{Nkb_c} P_{k,t}^{c} + \sum_{l=1}^{Lb} F_{l,t} \quad \forall b, \forall t
\tag{18}
$$

where, $F_{l,t}$ is power flow through line l in period t, ($F_{l,t} = \frac{1}{X_l}(\delta_{ls} - \delta_{lr})$), $\delta_{x,t}$ is voltage angle at node x in period t. The power flow through line l is limited as:

$$-F_{l,t}^{\min} \leq F_{l,t} \leq F_{l,t}^{\max} \quad \forall l, \forall t \tag{19}$$

- Real power generation constraints; The real power generated by DG units are constrained by (20) and (21) [21].

$$P_{i,t} \leq P_i^{\max} u_{i,t} - R_{i,t}^U \quad \forall i, \forall t \tag{20}$$

$$P_{i,t} \geq P_i^{\min} u_{i,t} + R_{i,t}^D \quad \forall i, \forall t \tag{21}$$

- Generation-side reserve limits; Constraints (22)–(24) impose limits on the provision of spinning reserve in terms of up and down regulations, as well as non-spinning reserve from the generating units.

$$0 \leq R_{i,t}^U \leq R_{i,t}^{U,\max} u_{i,t} \quad \forall i, \forall t \tag{22}$$

$$0 \leq R_{i,t}^D \leq R_{i,t}^{D,\max} u_{i,t} \quad \forall i, \forall t \tag{23}$$

$$0 \leq R_{i,t}^{NS} \leq R_{i,t}^{NS,\max}(1 - u_{i,t}) \quad \forall i, \forall t \tag{24}$$

- Demand-side reserve limits; Constraints (25) and (26) restrict the procurement of up and down reserves from the responsive loads.

$$0 \leq R_{j,t}^U \leq R_{j,t}^{U,\max} \quad \forall j, \forall t \tag{25}$$

$$0 \leq R_{j,t}^D \leq R_{j,t}^{D,\max} \quad \forall j, \forall t \tag{26}$$

- Unit commitment constraints; Equation (27) determines the start-up and shut-down status of units, while (28) states that a unit cannot start-up and shut-down during the same period [29].

$$y_{i,t} - z_{i,t} = u_{i,t} - u_{i,t-1} \quad \forall i, \forall t \tag{27}$$

$$y_{i,t} + z_{i,t} - 1 \leq 0 \quad \forall i, \forall t \tag{28}$$

- Generating units startup cost constraint; constraints (29) and (30) represent generating units startup cost limitations [21].

$$SUC_{i,t} \geq \lambda_{i,t}^{SU}(u_{i,t} - u_{i,t-1}) \quad \forall i, \forall t \tag{29}$$

$$SUC_{i,t} \geq 0 \quad \forall i, \forall t \tag{30}$$

The second-stage constraints account for stochastic operating conditions are the same as the first-stage constraints and mentioned in Appendix A.

4. Simulation Results and Discussion

4.1. Test Case

The simulations are performed over a modified residential MG which is presented in Figure 1 [30]. There are different types of distributed generation (DG) units in the MG including two micro-turbines (MT_1 & MT_2), two fuel cell (FC_1 & FC_2) units, and one gas engine (GE) unit. Also, there are a number of renewable-based prime movers in the system including three WTs, each with a capacity of 80 kW installed at bus 6, 9 and 16, respectively and two PV plants, each with a capacity of 70 kW installed at buses 5 and 10, respectively. The wind and PV power generation are a function of random wind speed and sun radiation, respectively. Their output power scenarios are reduced by applying a k-means

Appl. Sci. **2017**, *7*, 378

algorithm as shown in Figure 2. Technical specifications of the simulated MG components are given in Table 1 [34]. Moreover, the hourly load profile of the MG is illustrated in Figure 3 and is supposed to be divided into three different periods, namely valley period (00:00–5:00), off-peak periods (5:00–10:00, 16:00–19:00 and 22:00–24:00) and peak periods (11:00–15:00 and 20:00–22:00).

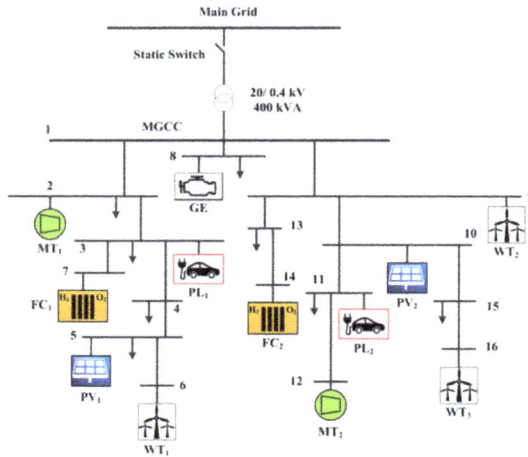

Figure 1. Single line diagram of the simulated microgrid MG.

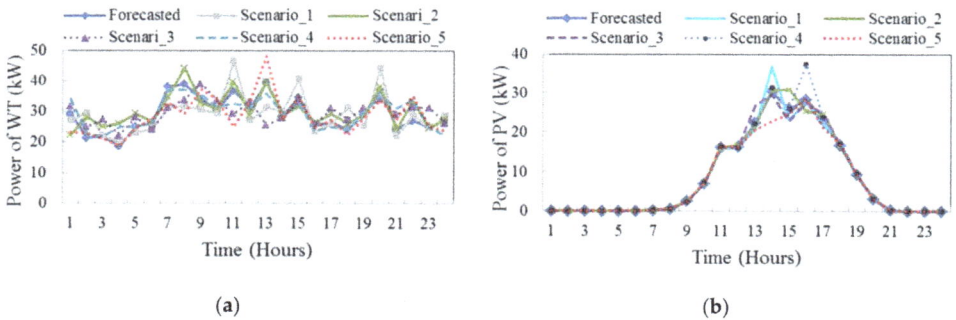

(a)   **(b)**

Figure 2. The output power of renewable energy resources (RESs) in the reduced generated scenarios, (**a**) each wind turbine (WT) and (**b**) each photovoltaic (PV) plant.

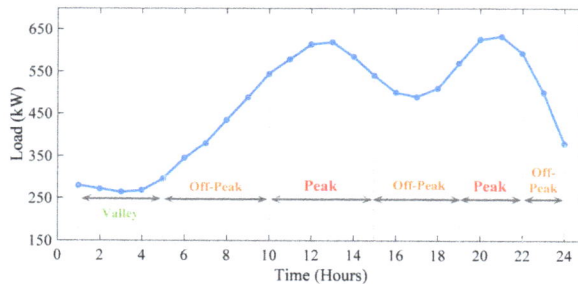

Figure 3. Total demand load curve of MG.

Table 1. Technical specifications of the simulated microgrid (MG) components.

Emission (kg/kWh)	$C^{R_{NS}}$ ($)	C^{R^D} ($)	C^{R^U} ($)	SDC ($)	SUC ($)	B ($)	A ($/kWh)	p^{max} (kW)	p^{min} (kW)	DG
0.550	0.019	0.020	0.021	0.080	0.090	0.043	0.851	150	25	MT_1
0.550	0.019	0.020	0.021	0.080	0.090	0.044	0.851	150	25	MT_2
0.377	0.015	0.015	0.015	0.090	0.160	0.028	2.552	100	20	FC_1
0.377	0.015	0.015	0.015	0.090	0.160	0.029	2.552	100	20	FC_2
0.890	0.017	0.017	0.017	0.080	0.120	0.031	2.120	150	35	GE
-	-	-	-	-	-	0.106	0	80	0	WT
-	-	-	-	-	-	0.548	0	70	0	PV

In this study, it is assumed that the total signed contracts for participating customers in DR programs are equal to 40% of the total load during the scheduling period. The price elasticity of demand is shown in Table 2, which is adopted from [27] with some modification. It is also assumed that there are two PLs with 40 charging stations in buses 3 and 11. The arrival time of EVs is modeled with a Gaussian distribution with $\mu = 19$ and $\delta^2 = 10$ [35]. Moreover, the EVs connected to the MG are assumed to be capable of providing slow, medium and fast charging modes [28,36]. For the studied MG, energy prices at different tariffs (RTP, TOU and CPP) are also depicted in Figure 4.

Table 2. Price elasticity of demand.

23–24	20–22	16–19	11–15	6–10	1–5	Hour
0.03	0.034	0.03	0.034	0.03	−0.08	1–5
0.03	0.04	0.03	0.04	−0.11	0.3	6–10
0.04	0.01	0.04	−0.19	0.04	0.034	11–15
0.03	0.04	−0.11	0.04	0.03	0.03	16–19
0.04	−0.19	0.03	0.01	0.04	0.034	20–22
−0.11	0.04	0.03	0.04	0.03	0.03	23–24

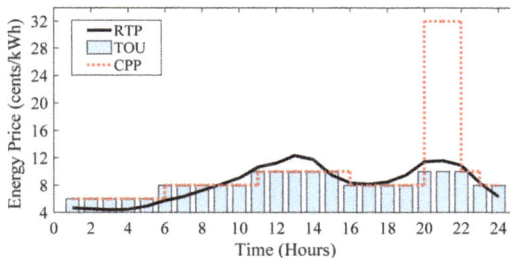

Figure 4. Energy prices at different tariffs (RTP, TOU and CPP).

The optimization horizon is considered to be a day with 24 time intervals. To simulate the environmental/behavioral uncertainties within the system, 3000 scenarios are generated based on Weibull, Beta, and Gaussian PDFs to represent different values for wind speed, irradiation and EVs owners' behaviors, respectively. In the next step, the k-means algorithm is applied to reduce the generated scenarios to an optimal subset that represents well enough the uncertainties. The reduced scenarios are then applied to the proposed mixed integer programming (MIP)-based optimization stage to minimize the expected cost at scheduling time horizon. The effect of demand-side participation in different TBR-based DR programs in the MG energy and reserve scheduling is also analyzed. The optimization is carried out by CPLEX solver using GAMS software (Release 24.7.3 r58181 WEX-WEI x86 64bit/MS Windows, TU Braunschweig, Braunschweig, Germany) [37] on a PC with 4 GB of RAM and Intel Core i7 @ 2.60 GHz processor (Intel, Santa Clara, CA, USA).

4.2. Presentation and Discussion of Results

We consider the following three cases for testing the effect of scheduling of DR programs and EVs on operation costs of the MG, load profile curve characteristics and profit of customers during the scheduling period.

- Case 1: without demand side participation and EVs commitment,
- Case 2: with demand side participation and without EVs commitment,
- Case 3: with demand side participation and EVs commitment.

It should be noted that Case 1 is considered to be a base case, so operating costs, load profiles and reserve scheduling in other cases are evaluated compared to the base case.

Case 1: In this case, DR programs are not considered and there is no contribution from EVs side. The scheduled energy and reserve capacity in this case is illustrated in Figure 5a,b, respectively. As shown in Figure 5a, based on the economic dispatch results, low-cost MT_1 and GE are used as base units to provide the energy. These generators are dispatched during the entire scheduling horizon to reduce the overall operating cost, while the other units (especially FC_1 and FC_2 due to their higher operating cost) are only dispatched at peak hour periods. As shown in Figure 5b, all the scheduled reserve capacity is provided by dispatchable DG units, including up-spinning reserve (Up/DGs), down-spinning reserve (Down/DGs), and non-spinning reserve (Non/DGs) in this case. Since the output power of WT and PV are intermittent, the required reserve power is provided by MTs, FCs and GE. It can be observed that when RESs power productions in scenarios are relatively different from the forecasted values, (i.e., in 10:00–14:00 and 19:00–22:00), more reserve capacity is scheduled accordingly. The total expected cost of MG operation as well as costs of providing energy and reserve services from DG units in case 1 are obtained as 897.833$, 436.622$ and 19.752$, respectively.

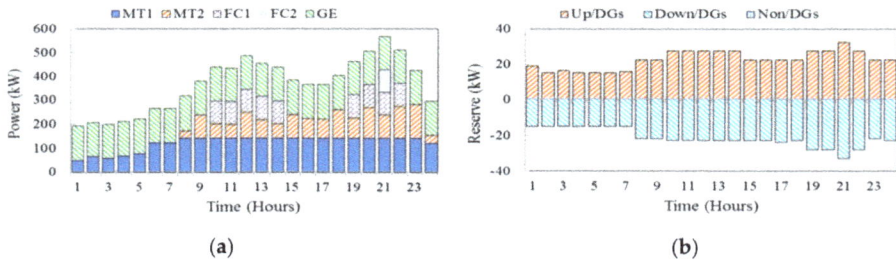

Figure 5. Hourly energy and reserves scheduling in case 1, (**a**) hourly energy and (**b**) reserve capacity.

Case 2: In this case, optimal operation of MG with demand-side participation (i.e., TBR-DR programs) but without EVs contribution is presented. The scheduled energy and reserve capacity in RTP programs are illustrated in Figure 6a,b, respectively. Comparing the results in Figures 5a and 6a demonstrates that with demand side participation in RTP schemes, the power provided by DG units is reduced at peak hours, specifically in 10:00–14:00 and 20:00–22:00. Likewise, during the hours with relatively high energy prices, the customers also reduce their consumption levels to save energy and get incentives. On the other hand, customers shift most of their consumptions into the time intervals with low energy prices, specifically in 01:00–05:00 (valley period, see Figure 3) to further reduce their running cost.

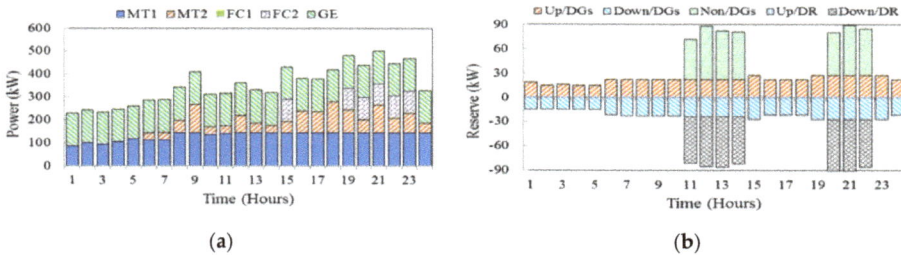

Figure 6. Hourly energy and reserves scheduling in case 2 under RTP programs (**a**) energy and (**b**) reserve.

As shown in Figure 6b, in this case, a part of the required reserve capacity is provided by demand side participation, including up-spinning reserve (Up/DR) and down-spinning reserve (Down/DR). Comparison of results in Figures 5b and 6b also shows that the participation of responsive loads can decrease the spinning reserve requirement of DG units and reduce the back-up energy costs. It can also be understood from the simulation results that the stochastic nature of wind and PV power generations, makes it necessary to allocate more reserve capacity to the time intervals (e.g., 10:00–14:00 and 19:00–22:00) when the risk of power shortage from RESs is higher. To this end, DR actions can provide a considerable portion of the needed upward reserve in the MG and decrease the MG operation cost. The expected operating cost of MG and the DGs energy and reserve costs in case 2 in RTP program are obtained as 872.943$, 416.789$ and 25.077$, respectively.

The scheduled energy and reserve capacity under TOU programs are illustrated in Figure 7a,b, respectively. By comparing the results in Figures 6a and 7a, it can be seen that the produced powers of DGs in TOU are slightly higher than ones in RTP programs at peak hours. This is due to the fact that during peak periods, participation of consumers in TOU programs is lower than that in RTP programs.

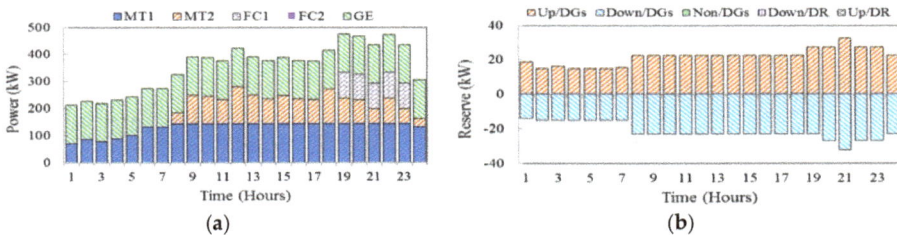

Figure 7. Hourly energy and reserves scheduling in case 2 under TOU programs (**a**) energy and (**b**) reserve.

It is also observed from Figure 7b that the scheduled reserves in these programs are different, especially in peak periods. This difference is due to the fact that the load reduction in TOU is less than that of in RTP at peak periods and the customers don't participate in downward reserve. Therefore, DGs non-spinning reserve scheduling is not required. The expected operating cost of MG and the DGs energy and reserve costs in case 2 under TOU programs are 881.164$, 420.549$ and 19.172$, respectively.

The hourly energy and reserve scheduling in case 2 in CPP programs are shown in Figure 8a,b, respectively. As mentioned before, the electricity price in CPP programs is the same as the price in TOU programs, except during hours 20:00 and 21:00. During these two hours, the price in CPP is four times greater than that of TOU. So, as Figure 8a shows, customers are highly encouraged to reduce their consumption as much as possible. As a result, the demand downward reserve increases in these two hours and consequently, the DGs non-spinning reserve increases. The expected operating cost of MG and the DGs energy and reserve costs are 850.395$, 410.590$ and 21.600$, respectively.

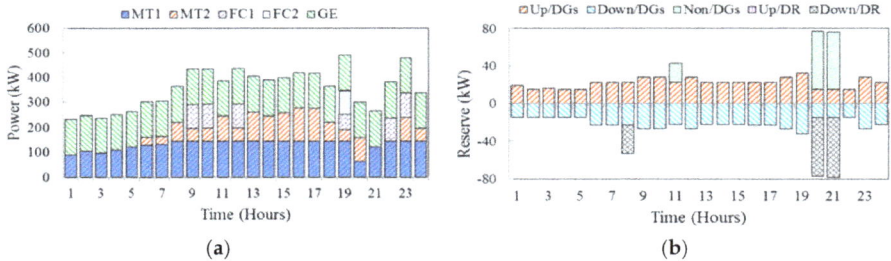

Figure 8. Hourly energy and reserves scheduling in case 2 under CPP programs (**a**) energy and (**b**) reserve.

Case 3: In this case, we evaluate the effectiveness of TBR-DR schemes in presence of EVs. In order to indicate the impact of different programs on responsive loads along with the presence of EVs, the same types of tariffs including RTP, TOU and CPP are implemented. Figure 9 illustrates the EVs daily charging and discharging power in different TBR- DR programs. As can be seen from the operating profiles, EVs are charged during low tariff hours (valley periods) and discharged during high tariff hours (peak periods). The energy and reserves scheduling in case 3 are shown in Figure 10a,b.

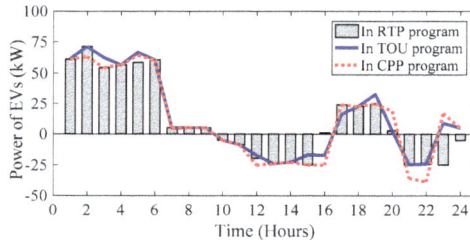

Figure 9. Charging and discharging power of EVs in TBR-based DR programs.

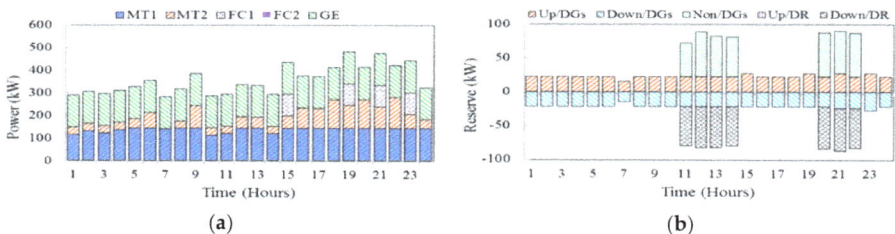

Figure 10. Hourly energy and reserves scheduling in case 3 in RTP programs (**a**) hourly energy and (**b**) Reserve.

Compared to Figure 6, it is observed that, the output power of generating units is flattened in the presence of EVs. It should be noted that, in case 3 the charging/discharging powers of EVs are added to the MG load which in turn affect the energy scheduling process; however, there is no effect on the reserve market as EVs are not considered on this occasion. So, it can be seen from Figure 10b that DR provides a part of the required reserve scheduling similar to case 2. The expected operating cost of MG and DGs energy and reserve costs in case 3 in RTP are obtained as 866.113$, 403.482$ and 23.482$, respectively, which are considerably lower than the corresponding values obtained in the previous cases. Thus, an efficient scheduling of EVs and responsive loads can improve the operation of the MG.

Figure 11a,b shows the energy and reserves scheduling in case 3 considering a TOU-based DR program. As can be seen from the results, during peak periods, EVs are discharged, and the operations of the costly units are delayed accordingly. So, in comparison with the two previous cases, the energy cost of DG units decreases. The expected operating cost of MG and DGs energy and reserve costs in case 3 considering a TOU scheme are obtained as 878.252$, 413.482$ and 21.084$, respectively.

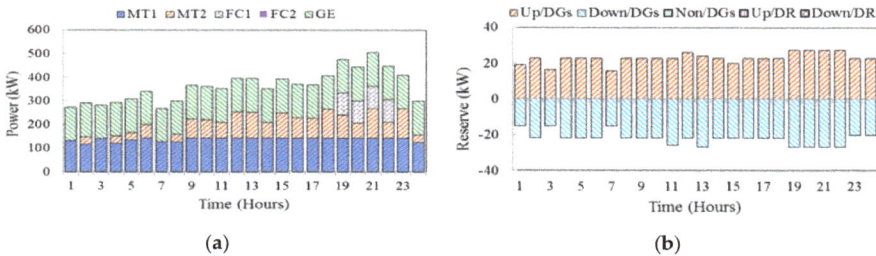

Figure 11. Hourly energy and reserves scheduling in case 3 in TOU programs (**a**) hourly energy and (**b**) Reserve.

The energy and reserves scheduling in case 3 regarding the CPP program are also shown in Figure 12a,b. Also, in this program, the participation of EVs in DR schemes decreases operating costs. The expected operating cost of MG and DGs energy and reserve cost values in this case study are obtained as 853.049$, 400.620$, and 20.801$, respectively.

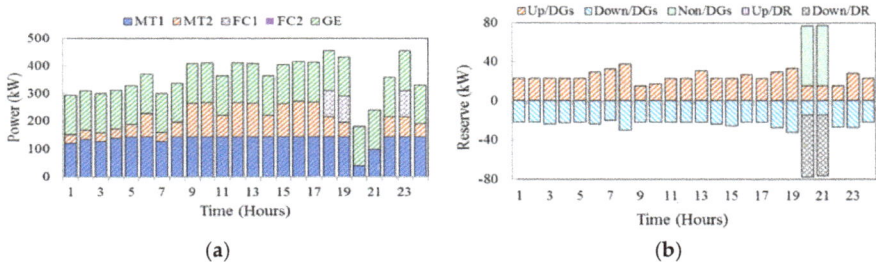

Figure 12. Hourly energy and reserves scheduling in case 3 in CPP programs (**a**) hourly energy and (**b**) Reserve.

The total load profile associated with the three cases in TBR programs are illustrated in Figure 13. As can be observed, with the application of DR programs, the total load decreases in peak periods when prices are high and increases in off-peak or valley periods when prices are relatively lower. This leads to smoother load profiles especially in cases 2 and 3. This load-shaping process can be better observed with active participation of EVs and their charging behavior during the valley period (when the price has its lowest value).

It can be also observed from Figure 13a,b, that the load profile in RTP and TOU schemes are relatively similar, but greatly different from the one in CPP scheme (Figure 13c) due to the price spikes at some time intervals and their effect on demand side participation.

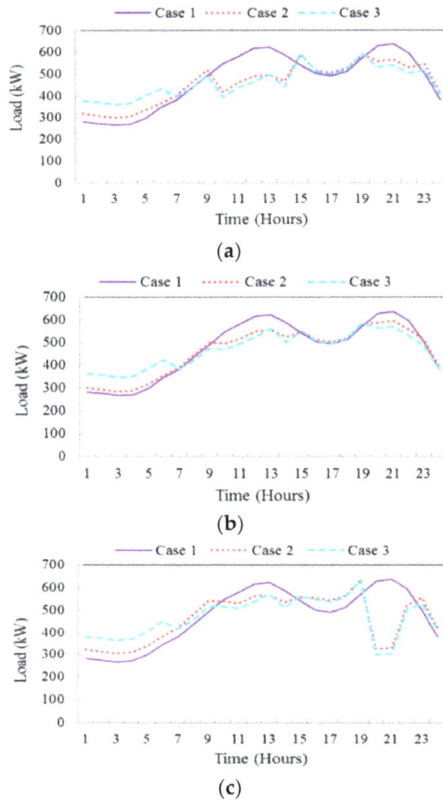

Figure 13. Daily load profile in three cases considering (**a**) RTP, (**b**) TOU and (**c**) CPP programs.

Table 3 compares the operational costs of the MG in different working conditions. The expected operating cost of MG, DGs and DR scheduled energy and reserve costs, start-up costs and start-down costs have been reported for the three cases. Comparison of results in cases 1 and 2 shows that the deployment of DR programs allows lower total operating cost to be obtained. The reason is that the expensive units are not dispatched to meet the demand of peak periods since peak loads are decreased due to the participation of responsive loads in different DR programs. Also, participation of both responsive loads and EVs in DR programs (case 3) can reduce the total operating cost more than the other cases where there are no DR action or EV support. In fact, in case 3, EVs discharging in peak hours (as shown in Figure 9) causes the decrement of expected cost of MG in comparison with cases 1 and 2. In other words, during peak hours, EVs discharge and supply peak loads; thus, the more expensive units may not be dispatched and, consequently, the energy cost of DGs is reduced. Moreover, in CPP programs, since customers are highly encouraged to reduce their consumption as much as possible at peak hours (at 19:00 and 20:00 as shown in Figure 13), the energy cost of DG units and as the result the total operating cost of MG has its lowest value. Moreover, in case 3, due to the uncertainty of EVs, the total cost of scheduling reserve of DGs and DR in each DR program is more than the one in case 2. However, it can be observed that the total deployed reserve cost of DGs and DR in cases 2 and 3 are almost the same, because deployed reserve is provided by DR and DG units and EVs do not participate, as they only affect the energy scheduling process.

Table 3. MG operating costs (in $) in three cases considering TBR-based DR programs.

Attribute	Case 1	Case 2			Case 3		
	No DR	RTP	TOU	CPP	RTP	TOU	CPP
Expected cost	897.833	872.943	881.164	850.395	866.113	878.253	853.049
Energy cost of DGs	436.622	416.790	420.549	410.590	403.482	413.482	400.620
Scheduling reserve cost of DGs	19.752	25.076	19.172	21.600	23.482	21.085	22.801
Scheduling reserve cost of DR	0	23.856	0	10.048	33.856	0	10.048
Energy cost of RESs	443.206	443.206	443.206	443.206	443.206	443.206	443.206
Deployed reserve cost of DGs	−2.365	10.182	−2.131	1.536	9.923	−2.081	1.189
Deployed reserve cost of DR	0	−56.785	0	−40.192	−56.785	0	−40.192
Start-up cost of DGs	0.78	0.78	0.62	0.87	0.87	0.64	1.02
Shut-down cost of DGs	0.27	0.27	0.18	0.35	0.35	0.25	0.46

In order to analyze the expected cost of MG with respect to load participation in DR, a sensitivity analysis is done and shown in Figure 14. With increasing customer participation in DR, the expected cost of MG is mitigated in all TBR-DR programs. As observed, in higher values of DR participation (i.e., more than 60%) the expected cost reduced slightly because in higher DR participants, new peaks of demand may occur (also known as rebound peak effect) and expensive units may need to be committed. Also, it is seen that in all the rates of DR participants, the expected cost in CPP program has lower value compared with other TBR-DR programs.

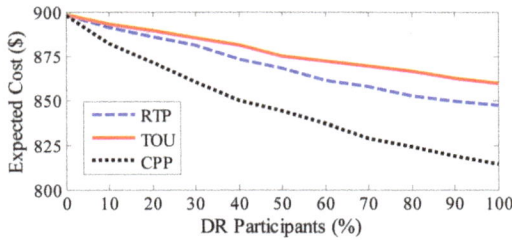

Figure 14. Expected cost of MG versus customers' participation in different TBR-DR programs.

5. Conclusions and Future Work

In this paper, the effect of the TBR-DR programs on reserve and energy scheduling in an isolated residential MG and in the presence of EVs were studied. A two-stage optimization model was developed to minimize the MG operation costs considering RESs and EVs uncertainties. The numerical results revealed that demand-side participation in energy and reserve scheduling reduces the total operating cost in different DR programs. The simulation results also demonstrated that in all TBR-DR programs, the participation of both responsive loads and EVs can reduce the energy cost of DGs and as the result the total operating cost of MG can decrease compared to the case where only DR actions are considered. Comparing the simulation results of TBR-DR programs also demonstrated that in CPP due to a great load reduction at peak price hours, the expected running cost of the system has its lowest value, and as a result, this program could be a proper alternative from the MG operator's viewpoint. In addition, the results showed that due to the uncertainty of EVs, the total cost of scheduling reserve of DGs and DR in each DR program is more than the one in case only with responsive loads. Moreover, it was shown that by increasing the participation of responsive loads in all TBR-DR programs (i.e., to more than 60%), the rate of decrement of expected cost may reduce due to the rebound peak problem.

Our future efforts will be mainly focused on developing an optimal scheduling model based on real-world uncertainties of DR resources and assessing their effects on islanded MG voltage and frequency security.

Appendix A

The second-stage constraints are as bellow:

Power balance equation in different scenarios: The active power balance in MG buses in each scenario is represented as follow [21]:

$$\sum_{i=1}^{Ngb} P_{i,t,s} + \sum_{w=1}^{Nwb} P_{w,t,s} + \sum_{p=1}^{Npb} P_{p,t,s} + \sum_{k=1}^{Nk_d} P_{k,t,s}^d + L_{b,t,s}^{shed} = L_{b,t,s} + \sum_{k=1}^{Nk_c} P_{k,t,s}^c + \sum_{l=1}^{Lb} F_{l,t,s} \quad \forall b, \forall t, \forall s \quad \text{(A1)}$$

where, $F_{l,t,s}$ is power flow through line l in period t and scenario s ($F_{l,t,s} = \frac{1}{X_l}(\delta_{n,t,s} - \delta_{r,t,s})$), which is limited as $-F_l^{min} \leq F_{l,t,s} \leq F_l^{max}$.

Generation-side reserve limits in each scenario [21]:

$$P_{i,t,s} \geq P_i^{min} u_{i,t,s} + R_{i,t,s}^D \quad \forall i, \forall t, \forall s \quad \text{(A2)}$$

$$P_{i,t,s} \geq P_i^{min} u_{i,t,s} + R_{i,t,s}^D \quad \forall i, \forall t, \forall s \quad \text{(A3)}$$

Deployed reserves limits from the generation-side: Constraints (A4)–(A6) enforce a limit on the procurement of up-, down- and non-spinning reserves from the generating units, respectively.

$$0 \leq r_{i,t,s}^U \leq R_{i,t,s}^U \quad \forall i, \forall t, \forall s \quad \text{(A4)}$$

$$0 \leq r_{i,t,s}^D \leq R_{i,t,s}^D \quad \forall i, \forall t, \forall s \quad \text{(A5)}$$

$$0 \leq r_{i,t,s}^{NS} \leq R_{i,t,s}^{NS} \quad \forall i, \forall t, \forall s \quad \text{(A6)}$$

Deployed reserves limits from the demand-side: Constraints (A7) and (A8) enforce a limit on the procurement of up- and down-spinning reserves from the responsive loads, respectively [29].

$$0 \leq r_{j,t,s}^U \leq R_{j,t,s}^U \quad \forall j, \forall t, \forall s \quad \text{(A7)}$$

$$0 \leq r_{j,t,s}^D \leq R_{j,t,s}^D \quad \forall j, \forall t, \forall s \quad \text{(A8)}$$

Involuntary load shedding: Equation (A9) represents the amount of inelastic load that can be shed by the MG operator in order to keep the system stable.

$$0 \leq L_{j,t,s}^{shed} \leq L_{j,t} \quad \forall j, \forall t, \forall s \quad \text{(A9)}$$

Decomposition of units power outputs; Constraint (A10) includes the scheduled day-ahead generation unit outputs with the deployed power in scenarios [21].

$$P_{i,t} = P_{i,t,s} + r_{i,t,s}^U + r_{i,t,s}^{NS} - r_{i,t,s}^D \quad \forall i, \forall t, \forall s \quad \text{(A10)}$$

Decomposition of demand consumption: The relationship between the amount of scheduled day-ahead responsive loads and up- and down-spinning reserves deployed in scenarios is represented by (A11) [21].

$$L_{j,t} = L_{j,t,s} - r_{j,t,s}^U + r_{j,t,s}^D \quad \forall j, \forall t, \forall s \quad \text{(A11)}$$

It should be noted that the up-reserves deployed by the demand-side is defined as a decrease in the consumption level, while down-reserve is defined oppositely.

Author Contributions: Mostafa Vahedipour-Dahraie and Amjad Anvari-Moghaddam developed the model; Mostafa Vahedipour-Dahraie simulated the case studies; Mostafa Vahedipour-Dahraie and Amjad Anvari-Moghaddam analyzed the data; Mostafa Vahedipour-Dahraie, Amjad Anvari-Moghaddam and Hamid Reza Najafi wrote the manuscript; Amjad Anvari-Moghaddam and Hamid Reza Najafi and Josep M. Guerrero provided their comments on the paper.

Nomenclature

Nb	Number of system buses.
Ng	Number of generating units.
Nj	Set of loads number.
Ns	Number of scenarios.
$Nw(Np)$	Number of WT (PV) units.
Nk	Number of EVs.
T	Scheduling time (24 h a day).
$i\ (j)$	Index of generating units (loads), running from 1 to $Ng(Nj)$.
b, n, r	Indices of system buses, running from 1 to Nb.
t	Index of time periods, running from 1 to T.
s	Index of scenarios, running from 1 to Ns.
$w\ (p)$	Index of WT (PV) units, running from 1 to $Nw(Np)$.
k	Index of EVs, running from 1 to Nk.
v	Wind speed (m/s).
$B(t)$	Customer's benefit in period t ($).
$BPR_t(SPR_t)$	Electricity baying (selling) price for EVs charging (discharging) in period t ($/kWh).
$C_{w,t}(C_{p,t})$	Energy bid submitted by WT w (PV p) in period t ($/kWh).
$C_{i,t}^{RU}(C_{i,t}^{RD})$	Bid of the up (down) -spinning reserve submitted by unit i in period t ($/kWh).
$C_{j,t}^{RU}(C_{j,t}^{RD})$	Bid of the up (down) -spinning reserve submitted by load j in period t ($/kWh).
$C_{i,t}^{RNS}$	Bid of the non-spinning reserve submitted by unit i in period t (cents/kWh).
$D(t)$	Power demand in period t (kW).
$D_{DR}(t)$	Power demand after implementing DR programs in period t (kW).
$E(t,t)$	Elasticity of load demand.
$SUC_{i,t}(SDC_{i,t})$	Start-up (Shut-down) cost of unit i in period t ($).
$\rho(t)$	Electricity price in period t ($/kW).
$P_{i,t}(P_{i,t,s})$	Scheduled power of unit i in period t (and scenario s) (kW).
$P_{w,t}(P_{w,t,s})$	Output power of WT w in period t (and scenario s) (kW).
$P_{p,t}(P_{p,t,s})$	Output power of PV p in period t (and scenario s) (kW).
$P_x^{max}(P_x^{min})$	Maximum (Minimum) generating capacity of unit x (kW).
$P_{k,t}^c(P_{k,t}^d)$	Charging (Discharging) power of EV k in period t (kW).
$P_{k,t}^{EV}$	Power of EV k in period t (kW).
$R_{i,t}^U(R_{j,t}^U)$	Scheduled up-spinning reserve for unit i (load j) in in period t (kW).
$R_{i,t}^D(R_{j,t}^D)$	Scheduled down-spinning reserve for unit i (load j) in period t (kW).
$R_{i,t}^{NS}$	Scheduled non-spinning reserve for unit i in period t (kW).
$r_{i,t,s}^U(r_{j,t,s}^U)$	Up-spinning reserve deployed by unit i (load j) in period t (and scenario s) (kW).
$r_{i,t,s}^D(r_{j,t,s}^D)$	Down-spinning reserve deployed by unit i (load j) in period t (and scenario s) (kW).
$S(t)$	Customer's income at period t ($).
V^{LOL}	Cost of involuntary load shedding for inelastic loads ($/kWh).
$L_{j,t}(L_{j,t,s})$	Power scheduled for load j in period t (and scenario s) (kW).
$L_{j,t,s}^{shed}$	Inelastic load shedding level of jth load in period t and scenario s (kW).
$F_{l,t}(F_{l,t,s})$	Power flow through line l in period t (and scenario s) (kW).
$\delta_{x,t}(\delta_{x,t,s})$	Voltage angle at node x in period t (and scenario s) (radian).
$\eta_c(\eta_d)$	Charging (Discharging) efficiency of EV
$u_{i,t}(u_{i,t,s})$	Binary variable, equal to 1 if unit i is scheduled to be committed in period t (and scenario s), otherwise 0.
$y_{i,t}(y_{i,t,s})$	Binary variable, equal to 1 if unit i is starting up in period t (and scenario s), otherwise 0.
$z_{i,t}(z_{i,t,s})$	Binary variable, equal to 1 if unit i is shut down in period t (and scenario s), otherwise 0.
$x_{k,t}$	Binary variable expressing the charging/discharging status of EV k, equal to 1 if it is charging, otherwise 0.

References

1. Khodaei, A.; Shahidehpour, M.; Bahramirad, S. A Survey on Demand Response Programs in Smart Grids: Pricing Methods and Optimization Algorithms. *IEEE Commun. Surv. Tutor.* **2015**, *17*, 564–571.
2. Anvari-Moghaddam, A.; Mokhtari, G.; Guerrero, J.M. Coordinated Demand Response and Distributed Generation Management in Residential Smart Microgrids. In *Energy Management of Distributed Generation Systems*; Lucian, M., Ed.; InTechOpen: Rijeka, Croatia, 2016; ISBN: 978-953-51-4708-4.
3. Shariatzadeh, F.; Mandal, P.; Srivastava, A.K. Demand response for sustainable energy systems: A review, application and implementation strategy. *Renew. Sustain. Energy Rev.* **2015**, *45*, 343–350. [CrossRef]
4. Mohagheghi, S.; Yang, F.; Falahati, B. Impact of demand response on distribution system reliability. In Proceedings of the IEEE PES General Meeting, San Diego, CA, USA, 24–28 July 2011.
5. Palensky, P.; Dietmar, D. Demand side management: Demand response, intelligent energy systems smart loads. *IEEE Trans. Ind. Inform.* **2011**, *7*, 381–388. [CrossRef]
6. Nguyen, D.T.; Negnevitsky, M.; de Groot, M. Pool-based demand response exchange-concept and modeling. *IEEE Trans. Power Syst.* **2011**, *26*, 1677–1685. [CrossRef]
7. Aghajani, G.R.; Shayanfar, H.A.; Shayeghi, H. Presenting a multi-objective generation scheduling model for pricing demand response rate in micro-grid energy management. *Energy Convers. Manag.* **2015**, *106*, 308–321. [CrossRef]
8. Jovanovic, R.; Bousselham, A.; Safak Bayram, I. Residential Demand Response Scheduling with Consideration of Consumer Preferences. *Appl. Sci.* **2016**, *6*, 16. [CrossRef]
9. Anvari-Moghaddam, A.; Rahimi-Kian, A.; Monsef, H. Optimal Smart Home Energy Management Considering Energy Saving and a Comfortable Lifestyle. *IEEE Trans. Smart Grid* **2015**, *6*, 324–332. [CrossRef]
10. Babar Rasheed, M.; Javaid, N.; Ahmad, A.; Ali Khan, Z.; Qasim, U.; Alrajeh, N. An Efficient Power Scheduling Scheme for Residential Load Management in Smart Homes. *Appl. Sci.* **2015**, *5*, 1134–1163. [CrossRef]
11. Rodriguez-Diaz, E.; Anvari-Moghaddam, A.; Vasquez, J.C.; Guerrero, J.M. Multi-Level Energy Management and Optimal Control of a Residential DC Microgrid. In Proceedings of the 35th IEEE International Conference Consumer Electronics (ICCE), Las Vegas, NV, USA, 8–11 January 2017.
12. Rabiee, A.; Sadeghi, M.; Aghaeic, J.; Heidari, A. Optimal operation of microgrids through simultaneous scheduling of electrical vehicles and responsive loads considering wind and PV units uncertainties. *Renew. Sustain. Energy Rev.* **2016**, *57*, 721–739. [CrossRef]
13. Alvarez, E.; Campos, A.M.; Arboleya, P.; Gutiérrez, A.J. Microgrid management with a quick response optimization algorithm for active power. *Int. J. Electr. Power Energy Syst.* **2012**, *43*, 465–473. [CrossRef]
14. Anvari-Moghaddam, A.; Seifi, A.R.; Niknam, T.; Alizadeh Pahlavani, M.R. Multi-objective operation management of a renewable MG (microgrid) with back-up micro-turbine/fuel cell/battery hybrid power source. *Energy* **2011**, *36*, 6490–6507. [CrossRef]
15. Anvari-Moghaddam, A.; Seifi, A.R.; Niknam, T. Multi-operation management of a typical microgrid using Particle Swarm Optimization: A comparative study. *Renew. Sustain. Energy Rev.* **2012**, *16*, 1268–1281. [CrossRef]
16. Chaouachi, A.; Kamel, R.M.; Andoulsi, R.; Nagasaka, K. Multiobjective intelligent energy management for a microgrid. *IEEE Trans. Ind. Electron.* **2013**, *60*, 1688–1699. [CrossRef]
17. Karangelos, E.; Bouffard, F. Towards full integration of demand-side resources in joint forward energy/reserve electricity markets. *IEEE Trans. Power Syst.* **2010**, *27*, 280–289. [CrossRef]
18. Shan, J.; Botterud, A.; Ryan, S.M. Impact of demand response on thermal generation investment with high wind penetration. *IEEE Trans. Smart Grid* **2013**, *4*, 2374–2383.
19. Peng, X.; Jirutitijaroen, P. A stochastic optimization formulation of unit commitment with reliability constraints. *IEEE Trans. Smart Grid* **2013**, *4*, 2200–2208.
20. Vrakopoulou, M.; Margellos, K.; Lygeros, J.; Andersson, G. A prob-abilistic framework for reserve scheduling and N-1 security assessment of systems with high wind power penetration. *IEEE Trans. Power Syst.* **2013**, *28*, 885–3896. [CrossRef]
21. Paterakis, N.G.; Erdinc, O.; Bakirtzis, A.G.; Catalão, J.P.S. Load-Following Reserves Procurement Considering Flexible Demand-Side Resources under High Wind Power Penetration. *IEEE Trans. Power Syst.* **2015**, *30*, 1337–1350. [CrossRef]

22. Aghaei, J.; Ahmadi, A.; Rabiee, A.; Agelidis, V.G.; Muttaqi, K.M.; Shayanfar, H. Uncer-tainty management in multiobjective hydro-thermal self-scheduling under emission considerations. *Appl. Soft Comput.* **2015**, *37*, 737–750. [CrossRef]

23. Amjady, N.; Aghaei, J.; Shayanfar, H.A. Stochastic multiobjective market clearing of joint energy and reserves auctions ensuring power system security. *IEEE Trans. Power Syst.* **2009**, *24*, 1841–1854. [CrossRef]

24. Dong, Q.; Yu, L.; Song, W.Z.; Tong, L.; Tang, S. Distributed demand and response algorithm for optimizing social-welfare in smart grid. In Proceedings of the 26th IEEE IPDPS, Shanghai, China, 21–25 May 2012; pp. 1228–1239.

25. Gholami, A.; Shekari, T.; Aminifar, F.; Shahidehpour, M. Microgrid Scheduling with Uncertainty: The Quest for Resilience. *IEEE Trans. Smart Grid* **2016**, *30*, 1337–1350. [CrossRef]

26. Kirschen, D.S.; Strbac, G. *Fundamentals of Power System Economics*; Wiley: Hoboken, NJ, USA, 2004.

27. Abdollahi, A.; Parsa-Moghaddam, M.; Rashidinejad, M.; Sheikh-El-Eslami, M.K. Investigation of Economic and Environmental-Driven Demand Response Measures Incorporating UC. *IEEE Trans. Smart Grid* **2012**, *3*, 12–25. [CrossRef]

28. Ma, Y.; Houghton, T.; Cruden, A.; Infield, D. Modeling the Benefits of Vehicle-to-Grid Technology to a Power System. *IEEE Trans. Power Syst.* **2012**, *27*, 1012–1020. [CrossRef]

29. Zakariazadeh, A.; Jadid, S.; Siano, P. Smart microgrid energy and reserve scheduling with demand response using stochastic optimization. *Electr. Power Energy Syst.* **2014**, *63*, 523–533. [CrossRef]

30. Rezaei, N.; Kalantar, M. Smart microgrid hierarchical frequency control ancillary service provision based on virtual inertia concept: An integrated demand response and droop controlled distributed generation framework. *Energy Convers. Manag.* **2015**, *92*, 287–301. [CrossRef]

31. Youcef, F.; Mefti, A.; Adane, A.; Bouroubi, M. Statistical analysis of solar measurements in Algeria using beta distributions. *Renew. Energy* **2002**, *26*, 47–67. [CrossRef]

32. Anvari-Moghaddam, A.; Monsef, H.; Rahimi-Kian, A.; Nance, H. Feasibility Study of a Novel Methodology for Solar Radiation Prediction on an Hourly Time Scale: A Case Study in Plymouth, UK. *J. Renew. Sustain. Energy* **2014**, *6*, 033107. [CrossRef]

33. Arthur, D.; Vassilvitskii, S. k-means++: The advantages of careful seeding. In Proceedings of the 18th Annual ACM-SIAM Symposium Discrete Algorithms (SODA '07), New Orleans, LA, USA, 7–9 January 2007; pp. 1027–1035.

34. Rezaei, N.; Kalantar, M. Stochastic frequency-security constrained energy and reserve management of an inverter interfaced islanded microgrid considering demand response programs. *Int. J. Electr. Power Energy Syst.* **2015**, *69*, 273–286. [CrossRef]

35. Rassaei, F.; Soh, W.S.; Chua, K.C. Demand Response for Residential Electric Vehicles with Random Usage Patterns in Smart Grids. *IEEE Trans. Sustain. Energy* **2015**, *6*, 1367–1376. [CrossRef]

36. Vahedipour-Dahraei, M.; Rashidizadeh-Kermani, H.; Najafi, H.R. A Proposed Strategy to Manage Charge/Discharge of EVs in a Microgrid Including Renewable Resources. In Proceedings of the 24th Iranian Conference on Electrical Engineering (ICEE), Shiraz, Iran, 10–12 May 2016; pp. 649–654.

37. The General Algebraic Modeling System (GAMS) Software. Available online: http://www.gams.com (accessed on 15 September 2016).

![applied sciences logo]

MDPI

Article

Optimal Scheduling of Industrial Task-Continuous Load Management for Smart Power Utilization

Jidong Wang *, Kaijie Fang, Jiaqiang Dai, Yuhao Yang and Yue Zhou

Key Laboratory of Smart Grid of Ministry of Education, Tianjin University, Tianjin 300072, China;
fangkaijie@tju.edu.cn (K.F.); daijiaqiang@gmail.com (J.D.); yyh8793@tju.edu.cn (Y.Y.); yuezhou@tju.edu.cn (Y.Z.)
* Correspondence: jidongwang@tju.edu.cn; Tel.: +86-136-5215-7798

Academic Editors: Josep M. Guerrero and Amjad Anvari-Moghaddam
Received: 9 November 2016; Accepted: 9 March 2017; Published: 14 March 2017

Abstract: In the context of climate change and energy crisis around the world, an increasing amount of attention has been paid to developing clean energy and improving energy efficiency. The penetration of distributed generation (DG) is increasing rapidly on the user's side of an increasingly intelligent power system. This paper proposes an optimization method for industrial task-continuous load management in which distributed generation (including photovoltaic systems and wind generation) and energy storage devices are both considered. To begin with, a model of distributed generation and an energy storage device are built. Then, subject to various constraints, an operation optimization problem is formulated to maximize user profit, renewable energy efficiency, and the local consumption of distributed generation. Finally, the effectiveness of the method is verified by comparing user profit under different power modes.

Keywords: demand response; distributed generation; smart power utilization; task-continuous load

1. Introduction

Faced with the increasingly severe circumstances of energy and the environment, many actions with respect to renewable energy have been taken around the world. One of the most important targets is to improve the access capacity of renewable energy on the user's side. With the development of smart grids and the implementation of related policies, many users have already attempted to use distributed energy. However, the asynchrony between the output of distributed generation and the users' loads leads to a low utilization rate of renewable energy and low profit for users. To make matters worse, the intermittence of renewable energy may lead to harmful effects on the distribution grid or even power failure due to the absence of a reasonable program [1]. Therefore, the local consumption of distributed generation is an important direction of energy management for both power grids and users. The users of electric power can be divided into residential users, utility users, commercial users, and industrial users. With the development of smart devices and advanced metering infrastructures, several programs of energy management for users have been proposed. The conception of user-side energy management has been proposed, including home energy management systems, building energy management systems, and enterprise energy management systems [2]. Based on user-side energy management systems, smart communities, smart industrial parks, and smart grids are steadily developing [3]. However, many existing energy management systems focus on energy monitoring systems rather than systems with optimization functions.

There is much literature about energy management and optimization. Among them, demand side management [4,5] and demand response technology are the main focus [6,7]. References [8–12] explore the method of improving energy efficiency with the assistance of demand response technology, aimed at water heaters, air conditioners, fridges, and washing machines. Reference [13] keeps the balance between economy and comfort by introducing comfort constraints in energy strategies that include distributed generation. In addition, the study of a micro-grid energy management system, analogous to the demand side energy management system, is also a hot topic for researchers [14]. As for the building energy management systems, in reference [15], a semi-centralized decision-making methodology using multi-agent systems was proposed to improve energy efficiency and reduce energy costs. Reference [16] introduces the actual experiment of using a building energy management system. After installing the building energy management system in a 21-floor building in Tirana, the total electrical energy footprint of the building was 135 kWh/m²/year; it was 200 kWh/m²/year before the installation of the system, which indeed is a massive drop. In addition, an enterprise energy management system was used in Guangzhou Iron and Steel Co., Ltd. (Guangzhou, China), to help them to arrange production [17].

However, there are still some issues that remain to be tackled. For one thing, the above existing studies, ignoring the distribution generation, cannot ensure the application of their technology to the demand-side energy management, which includes distributed generation. Moreover, the available research does not take the industrial task-continuous load into consideration when doing research on energy management. The main contribution of this paper lies in building a model of energy management considering the synchronization between the distributed generation and the industrial task-continuous load. In addition, this paper proposes a method that optimizes the industrial task-continuous load schedule to improve both the local consumption of the distributed generation and the economic benefit of users. Specifically, the contribution of this paper is as follows:

(1) The model of task-continuous load proposed in this paper is established, which can accurately describe the mathematical characteristic of the task-continuous load in the industry process.
(2) Regarding the states of industrial task-continuous load in time slots different from the controlled variables, the optimal solutions solved by this model of energy management can be directly applied to the industry process. That is to say, the method in this paper indeed has promising prospects of promotion in the industry.
(3) In addition, this paper shows that the model of energy management can shift the load to the period when the output of distributed generation is high and can shift the output power of the distributed generation into the battery. This results in a higher rate of self-occupied distributed generation, increasing the benefit for users.

2. The Model of Task-Continuous Load

Task-continuous load is a special kind of load that has been widely used in the industry process. Such kind of load is usually composed of several highly continuous devices. Each device is activated and works at different times, and the completion of the task requires the process flow containing all devices. Taking the process of the oxygen top blown converter, for instance, the processes that include making up raw material, adding molten iron, adding oxygen, and tapping are closely connected. In addition, the electricity consumption of each process is quite different from each other. Task-continuous load achieves cyclical fluctuation in terms of electricity consumption and often lasts for a long time, such as the loads of the production line and the iron metallurgy industry. The model of task-continuous load is established as follows.

It is assumed that k is the serial of the time slots, $k \in \{1, 2, 3, \ldots\ldots, T\}$, and $x_i(k)$ is the state of the ith device's switch at the kth time slot, where "1" refers to "on" while "0" refers to "off." P_i is the power rating of the ith process and $P_i(k)$ is the power of the ith process at the kth time slot. Therefore,

$$P_i(k) = P_i \times x_i(k) = \begin{cases} P_i & x_i(k) = 1 \\ 0 & x_i(k) = 0 \end{cases} \tag{1}$$

Task-continuous load should satisfy the following constraints.

2.1. The Constraint of On-Time

It is assumed that the on-time of the *i*th device is constrained by the working hours and process flow. That is to say, such a device must be turned off between the *a*th time slot and the *b*th time slot. Moreover, $k \in \{a, a+1, \ldots \ldots, b\} \subset \{1, 2, 3, \ldots \ldots, T\}$. This constraint is given by

$$\sum_{k=a}^{b} x_i(k) = 0 \tag{2}$$

2.2. The Constraint of Continuous Working

It is assumed that the *i*th device must work continuously from the *c*th time slot to the *d*th time slot. Moreover, $k \in \{c, c+1, \ldots \ldots, d\} \subset \{1, 2, 3, \ldots \ldots, T\}$. This constraint is given by

$$\prod_{k=c}^{d} x_i(k) = 1 \tag{3}$$

2.3. The Constraint of the Order of the Process Flow

It is assumed that, at the *e*th time slot, the *i*th procedure must be turned on, and the previous *j*th procedure must be accomplished. Moreover, $k \in \{1, 2, \ldots \ldots, e-1\} \subset \{1, 2, 3, \ldots \ldots, T\}$. If the *j*th procedure lasts for T_j time slots, this constraint is given by

$$x_i(k) \times \{\sum_{k=1}^{e-1} [x_j(k)/T_i] - 1\} = 0 \tag{4}$$

3. Model of Energy Optimization

User-side energy management, considering the type and significance of devices, provides a management scheme that makes optimal production plans to maximize users' benefits based on users' productive plans. Benefit maximization refers to the optimum of the property index such as time, cost, satisfaction, energy-saving, and emission reduction.

A typical user-side energy management system is shown in Figure 1. This system consists of a distributed generation unit, an energy storage unit, an inverter, a controller, and loads. In this system, electric supply is used to guarantee the normal operation of loads when the distributed generation unit fails to meet the demand.

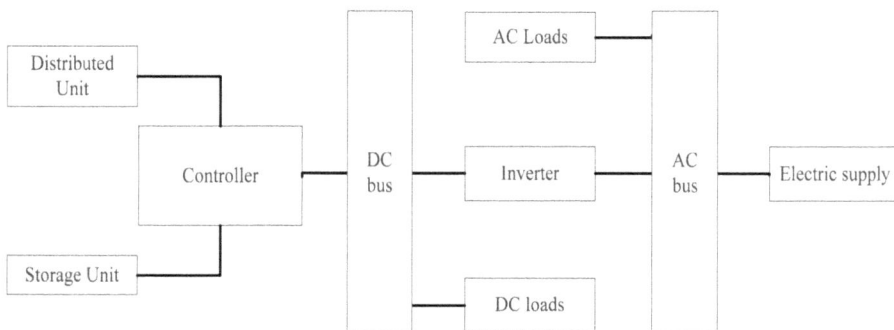

Figure 1. Structure of typical users' energy management system.

3.1. Maximum Revenue Target

For users with distributed generation units that are connected to the power grid, their economic benefit comes from two sources: (i) the subsidies for the local consumption of the distributed generation; and (ii) the money earned by selling redundant DG power to the power grid. In order to recover the cost of distributed generation equipment in the shortest time, the maximum revenue objective function ought to be adopted.

The maximum revenue target is shown as Equation (5):

$$
\begin{aligned}
\max F = & \sum_{k=1}^{T} C_{PV} P_{PV}(k) x_{PV}(k) \Delta t + \sum_{k=1}^{T} C_{Wind} P_{Wind}(k) x_{Wind}(k) \Delta t \\
& - \sum_{k=1}^{T} C_B P_B(k) \Delta t - \sum_{k=1}^{T} C_G(k) P_G(k) \Delta t
\end{aligned}
\tag{5}
$$

In Equation (5), the production planning cycle (usually be one day) is divided into T time slots, and each time slot lasts for Δt (usually one hour). In addition, the optimal strategy is allowed to execute at the initial moment of the kth time slot. The first part of users' revenue comes from the distributed generation (including the PV system and wind generation). Among the first part of the revenue, $x_{PV}(k)$ and $x_{Wind}(k)$ are the states of PV generation and wind generation, respectively, the controlled variables of the PV system and wind generation, where "1" refers to "on" and "0" refers to "off." $P_{PV}(k)$ and $P_{Wind}(k)$ represent the active power of PV generation and wind generation. At the same time, C_{PV} and C_{Wind} are the PV system's and wind generation's subsidized price of self-occupation, and usually are the same value. The second component of the objective is the revenue from the storage battery. Comparing with the distributed generation, the battery is allowed to charge or discharge. Therefore, the active power of the storage battery $P_B(k)$ can be negative. When $P_B(k) > 0$, the storage battery discharges, and, when $P_B(k) < 0$, the storage battery charges. Moreover, C_B is the generating cost of the storage battery. At last, the third part of the revenue comes from buying or selling the residuary power to the grid. Correspondingly, $P_G(k)$ represents the active power of the electric supply. When $P_G(k) < 0$, the distributed generation offers the electric energy back, and $C_G(k)$ is the difference between the acquisition price of DG and DG subsidized price of self-occupied at the kth time slot. However, when $P_G(k) > 0$, $C_G(k)$ is the electricity price of the initial moment of the kth time slot.

3.2. Active Power Balance Constraint

Active power balance constraint is shown as Equation (6):

$$
\sum P_L(k) = P_{ucl}(k) + \sum_{i=1}^{N} P_i(k) x_i(k) = P_{DG}(k) + P_B(k) + P_G(k)
\tag{6}
$$

$$
P_{DG}(k) = P_{PV}(k) x_{PV}(k) + P_{Wind}(k) x_{Wind}(k)
\tag{7}
$$

In Equation (6), N is the total number of task-continuous load. $P_i(k)$ is the active power of the ith task-continuous device at the kth time slot, and $x_i(k)$ is the state of the switch at the kth time slot, where "1" refers to "on" and "0" refers to "off." $P_{ucl}(k)$ represents the active power of other uncontrollable loads at the kth time slot. Therefore, $\sum P_L(k)$ is the whole power consumption at the kth time slot. In Equation (7), $P_{DG}(k)$ is total output of the distributed generation, which is directly supplied to the users' energy management system.

3.3. Storage Battery Constraint

Storage battery constraint is shown as Equations (8)–(10):

$$
\begin{aligned}
P_B(k) &> -P_{B,cmax} \\
P_B(k) &< P_{B,dmax}
\end{aligned}
\tag{8}
$$

$$SOC_{\min} < SOC(k) < SOC_{\max} \tag{9}$$

$$|SOC(1) - SOC(n)| < \delta \tag{10}$$

$P_{B,c\max}$ is the maximum charge power of the storage battery, and $P_{B,d\max}$ is the maximum discharge power of the storage battery. $SOC_{\max}$ and $SOC_{\min}$ are the superior limit and the inferior limit of the storage battery, respectively. $SOC_{\max}$ and $SOC_{\min}$ usually take 80% and 20%, respectively. $SOC(1)$ and $SOC(n)$ are the states of the battery at the start time and the terminal time. δ is usually 5%.

3.4. Transferable Load Contraint

To guarantee the production flow on the rails, the ability of the transferable task -continuous load ought to meet the demand of various kinds of production task constraints, such as the on-time constraint. The detailed computation is the same as Equations (1)–(4).

4. Model Solution

The model of energy management is like a knapsack problem: the independent variables are all integers, and the values of most independent variables are 0 or 1. The switch state of each load and the distributed generation is a question of whether a certain thing should be put into the bag or not. The constraints are analogous to the fact that the total weight of goods ought to be less than the capacity of the bag. The objective function is to maximize the total value of the goods in the bag. The whole model is to optimize the goods, and there is a total of 2^n solutions, where n is the number of the goods. Because the problem is linear, the dimension of the problem depends on the length of the schedulable period and the task of the schedulable load.

Therefore, in this paper, these switch states of each load and distributed generation are the controlled variables. The constraints of Equations (2)–(4) and (6)–(10) limit the feasible region of the controlled variables.

When the controlled variables of each solution are ascertained, the power of the task-continuous load, the distributed generation can be gained directly. While the power of the battery and grid should be calculated based on following the power supply flowchart, whose calculating process is shown in Figure 2. Firstly, the predicted power of the distributed generation $P_{DG}(k)$ and the total predicted power of the system $\sum P_L(k)$ should be calculated and compared so that further power flow between the users and the system can be ascertained. It is assumed that the power of the distributed generation and the total power of the system has been predicted. The method to account for inaccurate predictions is not discussed in this paper. Skipping to the second layer of the logic judgment, the state of the battery represented by its current $SOC(k)$ should be calculated. Finally, an economic scheduling assignment is provided based on (i) the state of the battery and the grid and (ii) the difference between the battery discharging cost and the price of the electric supply.

It is well known that a knapsack problem needs to be solved through iterative methods such as the branch and bound method, the dynamic planning method, and different kinds of intelligent algorithms. Generally, the PSO (Particle Swarm Optimization) algorithm, as one kind of intelligent optimization technique, has good performance in global optimization. Further, due to its simple structure, no need for gradient information of constraints, and few parameters in the algorithm, the PSO algorithm achieve suitable solutions in continuous and discrete optimization problems. Therefore, based on these features, the PSO algorithm is selected as the optimal method for solving.

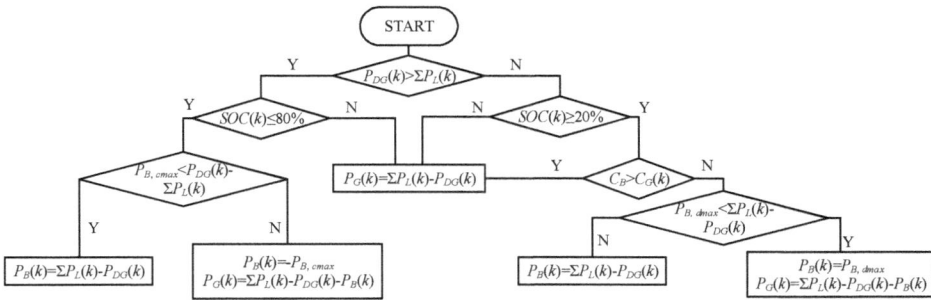

Figure 2. Flowchart of the power-supply-state judgment.

The PSO algorithm was firstly proposed by Kennedy and Eberhart [18]. This algorithm is a kind of stochastic evolutionary optimization method based on simulating bird population searches, that have good global search capacity. The basic theory of a PSO algorithm is briefly introduced as follows.

In the PSO algorithm, it is assumed that y_i and v_i are the position (the set of the controlled variables in this paper) and velocity (the evolutionary direction of the controlled variables) of the ith particle, while y_{i+1} and v_{i+1} represent the position and velocity of the next iteration. There are some parameters in the algorithm. For example, w is the inertia weight factor parameter in this algorithm, c_1 and c_2 are positive learning factor parameters, and r_1 and r_2 are random numbers between 0 and 1 and obey uniform distribution. Therefore, the position and velocity of the next iteration can be calculated by Equations (11) and (12):

$$v_j(i+1) = c_1 r_1 \left[y_{l,j} - y_j(i) \right] + c_2 r_2 \left[y_{g,j} - y_j(i) \right] + w v_j(i) \tag{11}$$

$$y_j(i+1) = y_j(i) + v_j(i+1). \tag{12}$$

If the position of a particle does not satisfy the constraints, the PSO algorithm should be modified and its velocity can be calculated by Equation (13):

$$v_j(i+1) = c_1 r_1 \left[y_{l,j} - y_j(i) \right] + c_2 r_2 \left[y_{g,j} - y_j(i) \right]. \tag{13}$$

In the process of the optimization of the modified PSO, the positions represent the states of the switch of each device in an industry process, and the velocities represent the changes in positions after each iteration [19].

The detailed optimization process of the PSO algorithm is shown in Figure 3. Firstly, the prediction of the load and distributed generation should be input into the algorithm. Then, the initial values of states of task-continuous load, PV generation, and wind generation (controlled variables) are obtained by randomly selecting in the set $\{0,1\}$. After attaining the initial values, it is easy to gain the power of the battery and grid though the power supply approach provided by Figure 2. Hence, the value of the objective function can be calculated by Equation (5).

Further, the PSO method is utilized to iterate by itself to generate a more optimal solution. Moreover, in every iteration, the constraints are requested to be checked. The iterative Equation (11) is replaced by Equation (13) when the constraints are not satisfied. After attaining a new solution, the algorithm goes back to the process of calculating the power of the battery and the grid. Moreover, the stop criterion of the PSO algorithm is whether the number of iterations reaches the setting value.

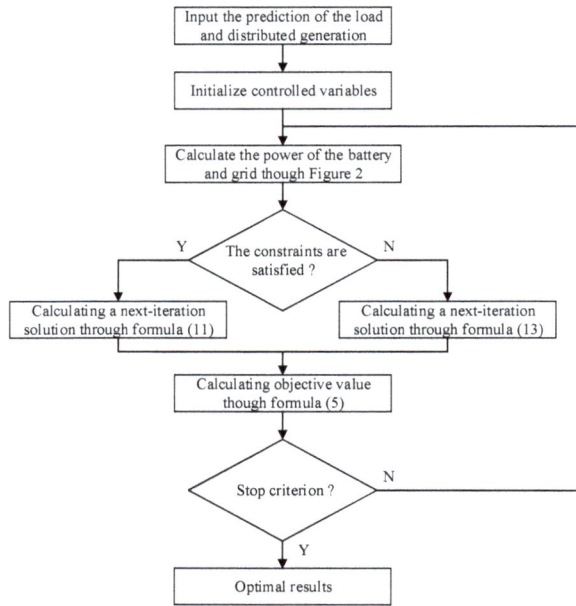

Figure 3. Flowchart of the PSO process.

5. Example Analysis

To simplify the model, the efficiency of the battery is regarded as 1. An example of industrial production processes in manufacturing devices is selected. It is assumed that this task includes 6 steps. In the production processes, Step 1 to Step 3 ought to be accomplished by order and Step 4 to Step 6 ought to be accomplished by order. That means before Step 2 starts, Step 1 should be accomplished already. Before Step 3 starts, Step 1 and Step 2 ought to be finished. Step 4, Step 5, and Step 6 are alike. The normal production work is divided into 24 time slots (each slot lasts for 1 h) and all the work ought to be carried out between 8:00 and 18:00.

As is shown in Figure 4, industrial electricity tariff is in accordance with city industrial production patterns, including three sections: peak, valley, and plain. In Figure 4, the unit is ¥. It represents RMB Yuan. The distributed unit is made up of a 600 kW PV system and a 300 kW wind-power generating unit. A 2000 Ah/480 V lead-acid storage battery is equipped in this system also. The life loss of the system is limited to 1/1000.

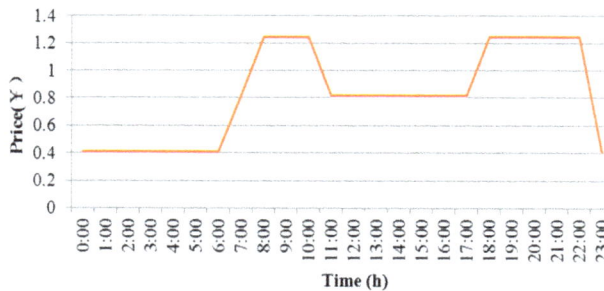

Figure 4. The price of electricity of a city.

Production time of the original, the original loads and the output of the distributed power are shown in Figure 5. The original loads consist of task-continuous processes and uncontrollable load, which is shown by the bars. Specifically, the load bars are stacked in a time slot. Therefore, the original maximum load power (800 kW) also occurs in the tariff peak.

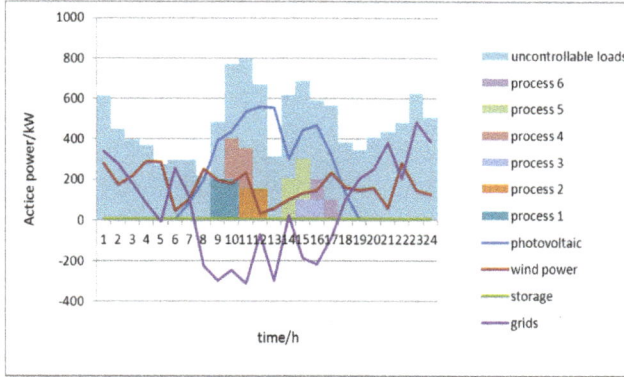

Figure 5. The stacked bars of the original loads and power curve of the distributed generation.

According to the energy management model proposed in this paper, the optimized power curve of the energy storage device and the optimal process order are shown in Figure 6. Comparing Figure 6 with Figure 4, it can be found that the period between 14:00 and 17:00 is the advantageous period where the electricity price is low and the output of PV maintains a high level. Therefore, Processes 1, 2, 4 and 5 are shifted into this period. Because of the characteristic of the task-continuous load, Processes 3 and 6 are shifted behind. At the same time, the results arrange the energy storage device to discharge to support Processes 3 and 6. The curve of grid in Figure 6 demonstrates that the schedule reduces both the selling power and the buying power in this period. The reduction of selling power benefits the local consumption of the distributed generation and the reduction of buying power is in favor of saving users' bills.

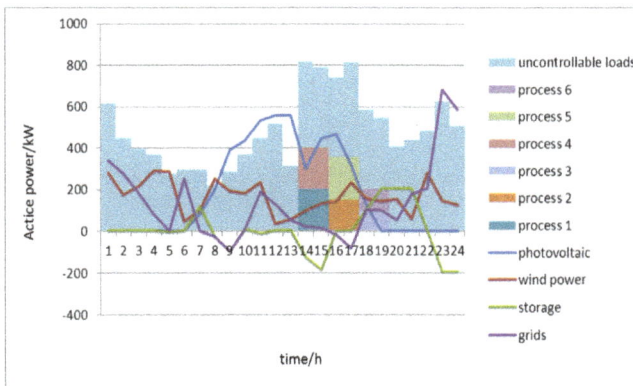

Figure 6. The process order and the power curve of the energy storage device after optimization.

Before optimization, the user's revenue is 6.53 Yuan. However, after optimization, the user's revenue is 8.57 Yuan. Therefore, the amount of revenue increase is 2.04 Yuan.

A comparison of production processes and power allocation sequences before and after optimization shows that this model reduces the load during the high tariff period time and transfers the load to the time slots when distributed energy generation is more prominent. From the power exchange with the grid and the actual power of the battery, we can verify that this model improves the local consumption of the distributed generation. On the other hand, the stability of the power grid has been improved with less distributed energy sent back to the grid. At the same time, the model improves the power curve and effectively responds to the electric tariff signal.

Specially, it is obvious that the former disordered working scheme of the task-continuous load is well reorganized and shifted. The optimal scheme of the task-continuous load shown in Figure 6 not only reduces the electricity bill but also makes the working process more compact and smooth. Due to the limitations of the time slot division and the user working time, the task-continuous load optimization becomes relatively concentrated. However, such planning has not uplifted peak loads, thus causing no extra pressure on the power grid. The result demonstrates the availability and efficiency of the energy management model proposed in this paper, which could be suitable for utilization in the industry.

6. Conclusions

Based on the task-continuous load characteristics, combined with the increasing integration of distributed generation, this paper presents a comprehensive consideration of distributed generation, energy storage charging/discharging, and task-continuous management strategies to optimize load operation. Compared to non-optimized energy management, the proposed strategy takes advantage of local distributed energy, improves the local consumption proportion of renewable energy sources, and improves the economic efficiency of users.

For user-side energy management issues, the toughest problem is not to obtain an optimal load schedule, but to accurately perceive and access users' needs. Therefore, future work will focus on how to analyze the influence of large amounts of load on the energy management system, how to improve optimization results, and how to realize a flexible and interactive smart power utilization in the environment of the smart grid.

Acknowledgments: The authors greatly acknowledge the support from the National Natural Science Foundation of China (NSFC) (51477111) and the National Key Research and Development Program of China (2016YFB0901102).

Author Contributions: Jidong Wang contributed to model establishing and paper writing. Many ideas on the paper are suggested by Kaijie Fang to support the work, and Fang did the job of performing the simulations. Jiaqiang Dai and Yuhao Yang analyzed the data. Yue Zhou reviewed the work and modified the paper. In general, all authors cooperated as hard as possible during all progress of the research.

Conflicts of Interest: The authors declare no conflicts of interest.

References

1. Elnozahy, M.S.; Salama, M.M.A. Technical impacts of grid-connected photovoltaic systems on electrical networks—A review. *J. Renew. Sustain. Energy* **2013**, *5*, 032702. [CrossRef]
2. Yang, W.; He, G.; Wang, W. Design and implementation of user energy management archetype system. *Autom. Electr. Power Syst.* **2012**, *36*, 74–79.
3. Fazel, A.; Sumner, M.; Johnson, C.M. Coordinated optimal dispatch of distributed energy resources within a smart energy community cell. In Proceedings of the International Conference on 2012 3rd IEEE PES Innovative Smart Grid Technologies Europe, Berlin, Germany, 14–17 October 2012; pp. 1–10.
4. Warren, P. A review of demand-side management policy in the UK. *Renew. Sustain. Energy Rev.* **2014**, *49*, 941–951. [CrossRef]
5. Shayesteh, E.; Moghaddam, M.P.; Yousefi, A.; Haghifam, M.R. A demand side approach for congestion management in competitive environment. *Eur. Trans. Electr. Power* **2013**, *20*, 470–490. [CrossRef]

6. Nayeripour, M.; Hoseintabar, M.; Niknam, T.; Adabi, J. Power management, dynamic modeling and control of wind/FC/battery-bank based hybrid power generation system for stand-alone application. *Eur. Trans. Electr. Power* **2012**, *22*, 271–293. [CrossRef]

7. Darby, S.J.; McKenna, E. Social implications of residential demand response in cool temperate climates. *Energy Policy* **2102**, *49*, 759–769. [CrossRef]

8. Zehi, M.A.; Bagriyanik, M. Demand Side Management by controlling refrigerators and its effects on consumers. *Energy Convers. Manag.* **2012**, *64*, 238–244.

9. Stadler, M.; Siddiqui, A.; Marnay, C.; Aki, H.; Lai, J. Control of greenhouse gas emissions by optimal DER technology investment and energy management in zero-net-energy buildings. *Eur. Trans. Electr. Power* **2011**, *21*, 1291–1309. [CrossRef]

10. Di Pimpinella, G.A. An event driven Smart Home Controller enabling L consumer economic saving and auto-mated Demand Side Management. *Appl. Energy* **2012**, *96*, 92–103.

11. Wang, C.; Zhou, Y.; Wang, J. A novel Traversal-and-Pruning algorithm for household load scheduling. *Appl. Energy* **2012**, *102*, 1430–1438. [CrossRef]

12. Pedrasa, M.A.A.; Spooner, T.D.; MacGill, I.F. Coordinated scheduling of residential distributed energy resources to optimize smart home energy services Smart Grid. *IEEE Trans. Smart Grid* **2010**, *1*, 134–143. [CrossRef]

13. Klein, L.; Kwak, J.; Kavulya, G. Coordinating occupant behavior for building energy and comfort management using multi-agent system. *Autom. Constr.* **2012**, *22*, 525–536. [CrossRef]

14. Mohammadi, M.; Hosseinian, S.H.; Gharehpetian, G.B. Optimization of hybrid solar energy sources/wind turbine systems integrated to utility grids as microgrid (MG) under pool/bilateral/hybrid electricity market using PSO. *Sol. Energy* **2012**, *82*, 112–125. [CrossRef]

15. Zhao, P.; Suryanarayanan, S.; Simoes, M.G. An energy management system for building structures using a multi-agent decision-making control methodology. *IEEE Trans. Ind. Appl.* **2013**, *49*, 322–330. [CrossRef]

16. Zavalani, O. Reducing energy in buildings by using energy management systems and alternative energy-saving systems. In Proceedings of the 2011 8th International Conference on the European Energy Market (EEM), Tirana, Albania, 25–27 May 2011; pp. 370–375.

17. Wang, Z.; Liu, X.; Wu, B. The study and application of steel enterprise energy management system. In Proceedings of the International Conference on Electric Information and Control Engineering (ICEICE), Guangzhou, China, 15–17 April 2011; pp. 4667–4670.

18. Eberhart, R.C.; Kennedy, J. A new optimizer using particle swarm theory. In Proceedings of the 6th International Symposium on Micro Machine and Human Science, Nagoya, Japan, 4–6 October 1995; Volume 1, pp. 39–43.

19. Sun, Z.; Li, L.; Bego, A.; Dababneh, F. Customer-side electricity load management for sustainable manufacturing systems utilizing combined heat and power generation system. *Int. J. Prod. Econ.* **2015**, *165*, 112–119. [CrossRef]

**applied
sciences**

MDPI

Article

Electrical Energy Forecasting and Optimal Allocation of ESS in a Hybrid Wind-Diesel Power System

Hai Lan [1], He Yin [1], Shuli Wen [1,*], Ying-Yi Hong [2], David C. Yu [3] and Lijun Zhang [1]

1 College of Automation, Harbin Engineering University, Harbin 150001, China; lanhai@hrbeu.edu.cn (H.L.); ccy1824@hrbeu.edu.cn (H.Y.); zhanglj7385@nwpu.edu.cn (L.Z.)
2 Department of Electrical Engineering, Chung Yuan Christian University, Chung Li District 320, Taoyuan City 32023, Taiwan; yyhong@ee.cycu.edu.tw
3 Department of Electrical Engineering and Computer Science, University of Wisconsin-Milwaukee, Milwaukee, WI 53211, USA; yu@uwm.edu
* Correspondence: wenshuli@hrbeu.edu.cn; Tel.: +86-451-82568560

Academic Editors: Josep M. Guerrero and Amjad Anvari-Moghaddam
Received: 4 November 2016; Accepted: 25 January 2017; Published: 14 February 2017

Abstract: Due to the increasingly serious energy crisis and environmental pollution problem, traditional fossil energy is gradually being replaced by renewable energy in recent years. However, the introduction of renewable energy into power systems will lead to large voltage fluctuations and high capital costs. To solve these problems, an energy storage system (ESS) is employed into a power system to reduce total costs and greenhouse gas emissions. Hence, this paper proposes a two-stage method based on a back-propagation neural network (BPNN) and hybrid multi-objective particle swarm optimization (HMOPSO) to determine the optimal placements and sizes of ESSs in a transmission system. Owing to the uncertainties of renewable energy, a BPNN is utilized to forecast the outputs of the wind power and load demand based on historic data in the city of Madison, USA. Furthermore, power-voltage (P-V) sensitivity analysis is conducted in this paper to improve the converge speed of the proposed algorithm, and continuous wind distribution is discretized by a three-point estimation method. The Institute of Electrical and Electronic Engineers (IEEE) 30-bus system is adopted to perform case studies. The simulation results of each case clearly demonstrate the necessity for optimal storage allocation and the efficiency of the proposed method.

Keywords: renewable energy; energy storage system; hybrid multi-objective particle swarm optimization; back-propagation neural network; power-voltage sensitivity analysis; three-point estimation method

1. Introduction

With the rapid development of renewable energy, interest in wind power has drawn more attention, as it possesses advantages such as free energy resources, non-greenhouse gas emission, and the ability of supporting rural areas. However, a high penetration of wind power raises a problem of system instability, caused by the nature of wind uncertainty. The integration of ESS is one of the best solutions to guarantee a stable power system with distributed wind resources [1].

An optimal allocation of ESSs in power systems can reduce total costs, enhance reliability and power quality, and, by determining the best locations and sizes of ESSs, improve voltage profiles [2]. Studies [3,4] show that an optimal planning of locations and sizes of ESSs in power systems can reduce a power system's costs and enhance a its reliability and power quality. A novel method has been presented in [5] for designing an energy storage system dedicated to the reduction of the uncertainty of short-term wind power forecasts up to 48 h. Wang et al. in [6] proposed a determination methodology for optimizing the capacity of an ESS that enables a wind power generator to meet the requirements of grid integration. To improve regulation effects, the segmentation method and automatic segmentation method are also applied to the proposed algorithm. An improved genetic algorithm is utilized in [7] to obtain the best energy savings and voltage profile by optimizing the location and size of ultra-capacitors. In [8], Xiao et al. proposed a capacity optimization method for a hybrid energy storage system taking SOC and efficiency into account. They used the maximal cumulative capacity and SOC constraint to calculate ESS capacity. Motaleb et al. [9] performed optimal sizing for a hybrid power system with wind and energy storage sources based on stochastic modeling of historical wind speed and load demand.

Since the uncertainties of wind power will lead to large errors in the results of optimally allocating ESSs, a power forecasting method is always used to predict power demand, spot price, and outputs of renewable energy and helps decision-makers determine an ESS's capacity more accurately. A large amount of literature [10–13] related to power prediction has been developed in recent years. The authors of [14–16] utilized artificial neural network (ANN) approaches for the optimum estimation and forecasting of renewable energy consumption by considering environmental and economic factors. Physical models were set up to predict solar irradiance and the output power of photovoltaic (PV) generation, which are based on numerical weather predictions and satellite images [17,18]. Bacher et al. developed a statistical approach based on a data-driven formulation using historical measured data to forecast renewable energy time series [19].

In addition to the determination of ESS capacity, it is also necessary to design the placement and rated power of energy storing devices, which is a complex parameter optimization problem. A multi-objective particle swarm optimization (MOPSO) algorithm has been widely utilized to solve the nonlinear, non-differentiable, multidimensional optimization problems [20,21]. Ganguly in [22] made use of MOPSO to plan the reactive power compensation of radial distribution networks with a unified power quality conditioner. Ramadan et al. [23] adopted the MOPSO technique to find the best capacitors in distribution systems that are connected to wind energy generations.

The focus of this paper is to optimally allocate ESSs in a power system integrated with wind power taking system costs, carbon dioxide emissions, and voltage fluctuations into account. In order to mitigate the influence of the uncertainties of wind power, a back-propagation neural network technique is utilized to forecast the power gap of the load and wind power using historical data. A hybrid multi-objective particle swarm optimization algorithm, which consists of a three-point estimation method, MOPSO, and a probabilistic power flow method, is developed to optimize the placements and sizes of the ESSs. Furthermore, power–voltage sensitivity analysis is proposed in this paper to select candidate buses for the installation of the ESS with the purpose of minimizing iteration times.

The novelty of this work, distinguishing it from previous studies, is as follows: (1) the candidate buses for the ESSs' installation were selected by the power–voltage (P-V) sensitivity analysis, which reduced the computational burden; (2) the capacity of the total ESS was determined by a forecasting method; (3) instead of running the Monte Carlo method, a three-point estimation method was employed to discretize the wind distribution; and (4) both the cost and greenhouse gas emissions were reduced by hybrid multi-objective particle swarm optimization.

The rest of this paper is organized as follows: Section 2 formulates the problem. Section 3 presents the method for solving it. Section 4 describes several case studies to demonstrate the proposed algorithm, and Section 5 draws conclusions.

2. Problem Formulation

2.1. Electrical Energy Forecasting

To ensure the secure and economic integration of wind turbines into a power system, accurate wind power and load demand forecasting has become critical of energy management systems [24]. In this paper, the back-propagation algorithm based on the artificial neural network is employed to fit the power curve of the difference between wind generation and load. The total capacity of the ESS is determined by the predicted wind power and loads.

Due to the uncertainties of renewable energy, it is difficult to predict the change in power and capacity of the ESSs. In this paper, the optimal storage capacity problem is formulated as a time series forecasting problem.

The outputs of the generation comprise two parts:

$$\begin{cases} P_d = P_s + P_{dev}^d \\ P_w = P_F^w + P_{dev}^w \end{cases} \tag{1}$$

Similarly, the load can be separated into two components as follows:

$$P_l = P_F^l + P_{dev}^l \tag{2}$$

Supposing that

$$P_s + P_F^w = P_F^l \tag{3}$$

The power difference between the actual power and the forecasting power is compensated by the ESS.

According to the requirement of a power system, the power balance in the hybrid wind/diesel/ESS system should be followed as Equation (4).

$$P_d + P_e + P_w = P_l \tag{4}$$

Consequently, the charging or discharging power of the ESS can be calculated by Equation (4) and the capacity of the ESS can be obtained herein, which is described by Equation (5).

$$P_e = P_l - P_d - P_w = \left(P_F^l + P_{dev}^l \right) - (P_s + P_{dev}^d) - (P_F^w + P_{dev}^w) \tag{5}$$

$$C_E = E_E^{\max}/u_{\min} = P_e^{\max} \times \Delta t / u_{\min} \tag{6}$$

Noted that the main effect of the ESS is to compensate for the power deviation from the forecast; as a result, the largest forecasting error is the total capacity of the ESSs.

2.2. Optimal Allocation of ESSs

The optimal placements and rated power of ESSs are formulated as a constrained nonlinear integer optimization problem where both the locations and sizes of the storage devices are discrete. The objective function encompasses the expected system costs, the emissions, and the voltage fluctuation under the consideration of multiple equality and inequality constraints.

2.2.1. Objective Function

The aims of this work are to minimize the total costs, greenhouse gas emissions, and voltage fluctuations by optimally determining the location and sizes of ESSs, while considering the uncertainties of the wind power generation. More specifically, the system costs contain the fuel cost of diesel generators, the operation cost of the WT, and the ESSs. The multi-objective functions are shown as follows:

$$\begin{cases} \min f1 = \sum_{i=1}^{3} Prob_i \cdot Cost_i \\ \min f2 = \sum_{i=1}^{3} Prob_i \cdot Emission_i \\ \min f3 = \sum_{i=1}^{3} Prob_i \cdot F_i \end{cases} \tag{7}$$

Notice that

$$Cost_i = \sum_{j=1}^{N} C(P_d) + C_w + C_e = \sum_{j=1}^{N} (a + b \cdot P_d + c \cdot P_d^2) + c^w \cdot P_w + c^e \cdot P_e \qquad (8)$$

$$Emission_i = \sum_{j=1}^{N} E(P_d) = \sum_{j=1}^{N} (d + e \cdot P_d + f \cdot P_d^2) \qquad (9)$$

$$F_i = \sum_{k=1}^{n} \left(\frac{V_k - V_k^{spec}}{\Delta V_k^{max}} \right)^2 \qquad (10)$$

where i is the scenario caused by the three-point estimate of wind power; V_k^{spec} is the expected voltage; ΔV_k^{max} is the maximum of voltage deviation. The expected value of Equation (7) is to calculate the desired system cost and the emissions by optimally allocating the ESSs and by determining the outputs of all the different types of generators factoring in the wind distribution. However, the voltage will fluctuate sharply with the change in the wind power generation, so the voltage profile is improved by the third objective function of (7).

2.2.2. Constraints

To a hybrid wind/diesel/ESS power system, the following operational constraints should be satisfied.

1. Equality Constraints: the power balance that is related to the nonlinear power flow equations is considered in this paper, which is shown in Equation (11).

$$\begin{cases} P_i - V_i \sum\limits_{j=1}^{N} V_i (G_{ij} \cos \delta_{ij} + B_{ij} \sin \delta_{ij}) = 0 \\ Q_i - V_i \sum\limits_{j=1}^{N} V_i (G_{ij} \sin \delta_{ij} - B_{ij} \cos \delta_{ij}) = 0 \end{cases} \qquad (11)$$

2. Inequality Constraints: the inequality constraints are those associated with the bus voltage V_k, the reactive power of generation Q_{Gi}, the tap of the transformer T_i, and the maximum charge/discharge power of the ESS.

$$\begin{cases} V_{min} \leq V_k \leq V_{max} \\ T_{min} \leq T_i \leq T_{max} \\ Q_{Gmin} \leq Q_{Gi} \leq Q_{Gmax} \\ -n_c \cdot C_E \leq P_e \leq n_d \cdot C_E \end{cases} \qquad (12)$$

where n_c and n_d are taken to be 3 C in this paper.

3. Solution Method

3.1. Discretizing Wind Distribution

The optimal allocation of the ESSs is always determined in the worst case (peak load without wind power) or the historical time series of the power [25,26]. The goal of this paper is to obtain the sitting and sizing of the ESSs in a more convenient way. Instead of using the Monte Carlo method, a three-point estimation method is adopted to discretize the distribution of wind power.

3.1.1. Wind Distribution

The probabilistic description of wind speed can be accurately presented by the Weibull distribution [27]. Due to the great applicability, Weibull distribution has been widely used to describe the probably distribution of the wind speed, which is defined as follows:

$$f(x; \lambda, k) = \frac{k}{\lambda} \left(\frac{x}{\lambda}\right)^{k-1} e^{-(x/\lambda)^k} \tag{13}$$

where k represents the shape parameter, and λ is the scale parameter.

To obtain the wind power distribution, a linear approximation equation that established the relationship between the wind speed and the wind power is presented in (14):

$$Y = \begin{cases} 0 & if \ x \le V_{ci} \ or \ x > V_{co} \\ \alpha + \beta x & if \ V_{ci} \le x \le V_{no} \\ M & if \ V_{no} \le x \le V_{co} \end{cases} \tag{14}$$

where M is the maximum power of wind turbine; α and β are the linear coefficients; and V_{ci}, V_{co}, and V_{no} denote the cut-in wind speed, the cut-out wind speed, and the normal wind speed, respectively.

3.1.2. Discretizing Wind Speed Distribution

The aim of the discretization scheme is to group the values of the continuous random variable into a three finite group.

The continuous sequence of wind speed can be discretized into three points, as shown in the following formula:

$$y_i = \mu_x + z_i \cdot \sigma_x. \tag{15}$$

Notice that

$$\begin{cases} z_i = \frac{\lambda_3}{2} + (-1)^{3-i} \sqrt{\lambda_4 - \frac{3\lambda_3{}^2}{4}}, i = 1, 2 \\ z_3 = 0 \end{cases} \tag{16}$$

where y_i is the discrete wind speed

Define

$$\begin{cases} \mu_x = \int\limits_0^M x f(x; \lambda, k) dx \\ \sigma_x = \int\limits_0^M (x - \mu_x)^2 f(x; \lambda, k) dx \\ \lambda_i = \int\limits_0^M \left(\frac{x - \mu_x}{\sigma_x}\right)^i f(x; \lambda, k) dx \end{cases} \tag{17}$$

where μ_x is the mean value of x; σ_x is the standard deviation of x; λ_i is the ith central moment of x.

Notice that

$$\begin{cases} P_i = \frac{(-1)^{3-i}}{z_i(z_1 - z_2)}, i = 1, 2 \\ P_3 = 1 - P_1 - P_2 \end{cases} \tag{18}$$

where P_i is the probability corresponding to y_i.

A ten-year daily wind speed data for the city of Madison, USA, was utilized to fit the Weibull distribution. Specifically, the rated power for wind generation was 113 MW with the total maximum load of 283 MW, and the cut-in, cut-out, and normal speed of wind turbines was 3.5 m/s, 40 m/s, and 13.5 m/s, respectively. The three-point discrete distribution was calculated and is shown in Figure 1.

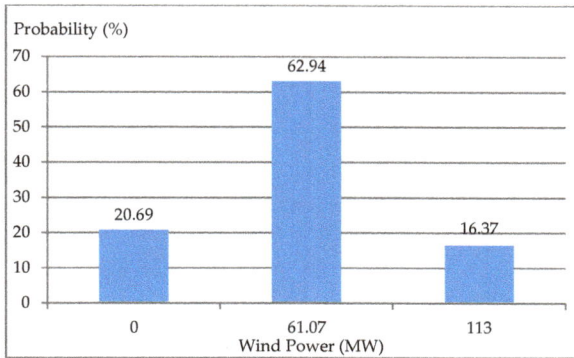

Figure 1. Three-point discrete distribution of wind power.

3.2. BPNN Prediction

A standard BPNN is a multilayer feed-forward neural network with error backward propagation, including an input layer, one and more hidden layers, and an output layer. Each layer is composed of a number of neurons that are connected by weights and thresholds. To solve a complicated problem, a complex neural network structure must be established by increasing the number of neurons and layers, and the structure of the network should match the problem [28]. Figure 2 illustrates a three-layer BPNN with a sigmoid function for the hidden layer and a linear function for the output layer.

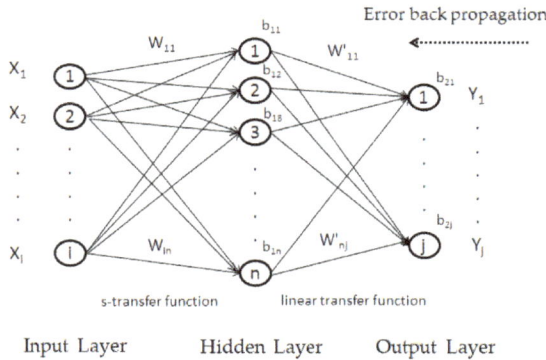

Figure 2. Basic structure of a back-propagation network.

During the training progress, an input pattern is given to the input layer of the network. Based on the given input pattern, the network will compute the output in the output layer. This network output is then compared with the desired output pattern. The aim of the back-propagation learning rule is to define a method of adjusting the weights of the networks. Eventually, the network will give the output that matches the desired output pattern given any input pattern in the training set.

In this paper, a dynamic artificial neural network integrated with the non-linear auto-regressive model with exogenous inputs (NARX) [29,30] is employed to forecast the output power of the generation, which contains an input layer, a hidden layer with delays, and an output layer with 2, 10, and 1 neurons, respectively.

As can be seen from Figure 3, a non-linear autoregressive with exogenous inputs dynamic neural network model establishes a relationship between the optimal scheduling of diesel generation and the actual data of wind power and the load, and is detailed in Equation (19).

$$P_{diesel}(t) = f[P_{diesel}(t-1), \ldots, P_{diesel}(t-d); P_{wind}(t-1), \ldots, P_{wind}(t-d)] + \varepsilon(t), \, d \geq 1 \quad (19)$$

where d is the delay-order of independent variable; $P_{wind}(t)$ is *wind* power at t hour; $P_{diesel}(t)$ is the output power of the diesel generation power in one-hour ahead scheduling. It should be noticed that, in the NARX model, load demand is not the independent variable because it has been contained in $P_{diesel}(t)$.

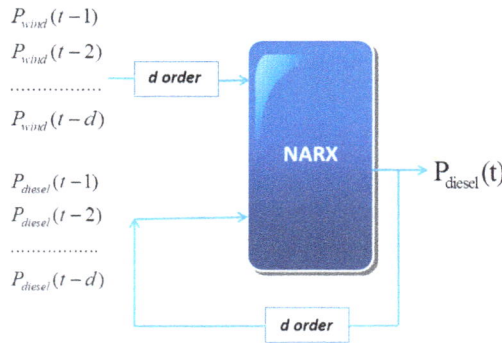

Figure 3. Non-linear auto-regressive model with exogenous inputs—back-propagation neural network (NARX-BPNN).

3.3. HMOPSO

Particle swarm optimization (PSO) is an intelligent optimization technique that was firstly put forward in 1995 [31–33], and the fundamental idea of the PSO algorithm is to randomly generate particles in the solution space and make each particle gradually approach the optimal solution [34,35]. In this paper, a hybrid MOPSO algorithm integrated with a non-dominated sorting genetic algorithm (NSGA-II) is programmed by MATLAB (Version R2010b, the MathWorks, Natick, MA, USA, 2010) to optimize the placements and sizes of ESSs in a hybrid wind/diesel power system. MOPSO is utilized to update the position and velocity for each particle to search the best allocation of the ESS, and NSGA-II is used to find a substantially improved spread of solutions and an improved convergence [36]. The procedure of the proposed method can be summarized as follows:

1. Forecast the difference between wind power and load demand and calculate the total capacity of the ESSs.
2. Randomly generate a population with a certain number of particles for initializing all generators' voltage, the output power, and the position and size of the ESSs. The random selections of the swarm of particles considering constraints and corresponding velocity for each particle are initialized.
3. Discretize the joint wind power distribution into a three-point distribution by the proposed estimation method, which is discussed in Section 3.1.
4. Select the candidate buses for installing ESSs via *P-V* sensitivity analysis.
5. Through probabilistic power flow, evaluate the particles by fitness function and recall their best positions associated with the best fitness value.
6. Check and preserve the *pbest* (particle best) and *gbest* (global best); if the algorithm has not yet found the minimum cost, emissions, and voltage fluctuations, update the *pbest* and *gbest*.

7. Duplicate the initial population to another population to form a combined population and update the position and velocity of each particle.
8. Sort the members in the new population through NSGA-II with an elitism algorithm for selecting the best solutions to renew the original population.
9. Repeat Steps 5–8 until all scenarios are considered.

4. Results and Discussion

4.1. Electrical Energy Forecasting

One-year wind power and load historical data, which are shown in Figure 4, of the hybrid system are used to predict the output power of the diesel generation.

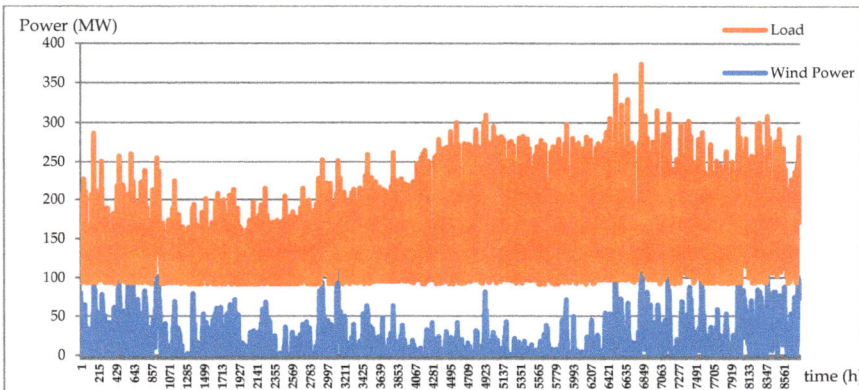

Figure 4. One-year data of wind generation and load.

In this paper, a sodium–sulfur battery is used as the energy storage unit and the slowest battery charge/discharge rate is 0.5 C. Three types of forecasting methods are studied and conducted to calculate the time series of the outputs of the generations. The total capacity of the ESSs is determined herein. The performances of the predicting methods are compared, and the results are depicted in Table 1.

Table 1. Energy storage system (ESS) capacity based on different forecasting methods.

Method	Network Type	Capacity (MWh)	Maximum Deviation	Mean Deviation	Variance
No prediction	-	1465.68	100%	21.34%	0.00686
Persistence model	-	226	52.09%	8.58%	0.009
Static prediction	Standard BP	217.48	50.13%	5.28%	0.00695
	Variable gradient BP	213.56	49.23%	5.30%	0.00634
	BFGS	205.84	47.45%	5.13%	0.00562
	Conjugate gradient BP	209.57	48.31%	5.02%	0.00533
	LM	201.07	46.35%	4.72%	0.00484
Dynamic prediction	NAR BP	56.52	13.03%	0.86%	0.00022
	NARX BP	54.4	12.54%	0.89%	0.00019

It can be seen from Table 1 that, with the help of the NARX-BPNN method, the total capacity of the ESS is reduced from 1465.68 MWh to 54.4 MWh with a minimum variance of 0.00019. Compared with the other prediction model, not only the maximum deviation but also the mean deviation of the dynamic prediction method achieves the lowest, which is nearly 4 times less than that of the persistence model. Figure 5 presents the performance of the NARX-BPNN method.

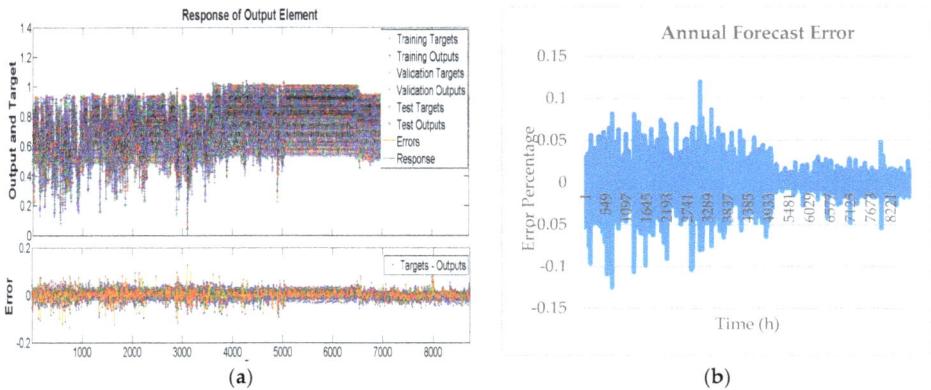

Figure 5. (**a**) Simulation result of NARX-BPNN; (**b**) Annual forecast error of NARX-BPNN.

4.2. Sensitivity Analysis

Power vs. voltage characteristics, known as the *P-V* curve, has been widely used as a complementary analytical tool to the dynamic study by many utilities. It depicts the loading and generating capability of each bus in the power network with respect to voltage stability. The *P-V* curve enables system planners and operators to reduce the risks of systems accidentally entering unstable regions [37].

In the paper, in order to minimize the possible placements of energy storage systems, thereby reducing the computational complexity of the PSO algorithm, *P-V* sensitivity analysis is conducted. The IEEE 30-bus system is selected to verify the ability performance of the proposed algorithm. The system consists of five generations and 20 loads, where Bus 1 is the slack bus, Bus 2 is installed with wind generation rated as 113 MW, Buses 5, 8 11, and 13 are defined as PV nodes, and other buses are PQ nodes [38]. Wind generation is added to Bus 2 rated as 113 MW. In order to simulate the extreme actual severe operating condition and expose the weaknesses and limitations of the system, the lengths of branches from Bus 1 to Bus 3 and Bus 1 to Bus 2 are doubled.

Buses that have more variation according to the change of loading and/or generating conditions are identified as sensitive buses. By placing energy storage systems at sensitive buses, the voltage profiles not only at these sensitive buses but also at adjacent buses will be improved. Therefore, by selecting proper locations for energy storage systems, the overall system voltage profiles can be improved, and system costs and losses thus minimized.

Figure 6 depicts *P-V* curves of generation and load areas in the IEEE 30-bus system. Buses 7, 10, and 12 in the generation area shown in Figure 6a, and Buses 25, 26, 28, the 30 in the load area shown in Figure 6b are noticeably more sensitive than other buses; in other words, they are able to contribute more voltage stability than other buses when energy storage devices are installed. It is worth noticing that Buses 1, 5, 8, 11, and 13 are not considered to install ESSs, as they are either connected to the bulk grid or already installed generations.

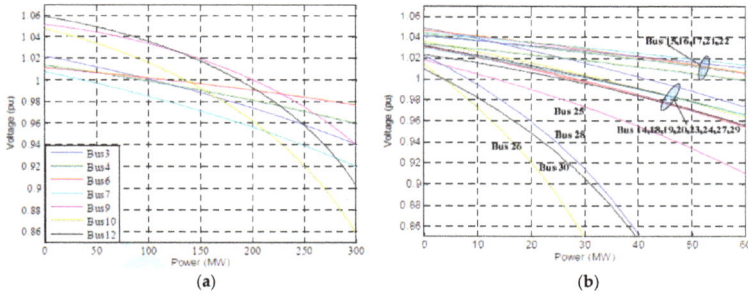

Figure 6. (a) Power–voltage (*P-V*) curves for generation areas of the IEEE 30 bus system; (b) *P-V* curve for loading areas of the Institute of Electrical and Electronic Engineers (IEEE) 30 bus system.

4.3. Economic Analysis

Considering the actual situation, the impacts of integration of the wind power and ESS into a transmission system are studied and compared in three cases in consideration of three scenarios of wind power, which are obtained from 3-point estimation method (3-PEM) to illustrate the effectiveness of the HMOPSO method. Table 2 presents the fuel cost parameters for each diesel engine, and the operation cost parameters c^w, c^e are 45 \$/MW and 5.8 \$/MW, respectively.

Table 2. Fuel cost parameters of diesel engines.

Generator	a	b	c
Gen 1	0	20	0.038432
Gen 2	0	40	0.01
Gen 3	0	40	0.01
Gen 4	0	40	0.01
Gen 5	0	40	0.01

Case 1: A probabilistic load flow analysis for the peak load condition in the IEEE 30-bus system without ESS installation;

Case 2: A cost study with an ESS completely installed in Bus 2 where wind turbine is located;

Case 3: HMOPSO considering seven sensitive nodes as candidate locations.

Figure 7 reveals the economic performance of optimization results in three cases.

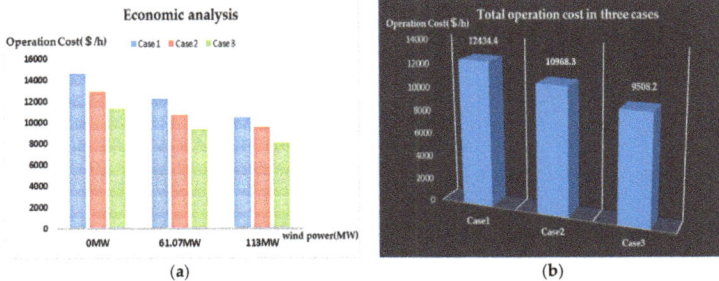

Figure 7. (a) Operation cost with consideration of wind distribution; (b) Total operation cost in three cases.

It can be seen from above that the total operation cost is reduced with the changes of the wind power from 0 MW to 113 MW. In Case 1, the load demand is only supplied by the diesel generation such that the system has to face a high cost of $12,434.4/h. Even though the ESS is introduced to Case 2, the system cost is still high which implies that the distributed ESSs and optimization method is necessary. More specifically, the costs in Case 2 are $12,877, $10,721, and $9507 with the change of the wind power from 0 MW to 113 MW. Compared with other cases, the cost in Case 3 is the lowest with $11,270/h, $9312/h, and $8036/h, respectively corresponding to the three different wind power outputs. If the system operate in one year (8760 h), the system will save $25,633,512.

4.4. Carbon Emission Analysis

Table 3 presents the carbon emission parameters of diesel engines [39]. Figure 8 reveals the benefit from the renewable energy. It shows that total carbon emission decreased from 17,064 kg/h to 15,364 kg/h with increasing wind power in Case 1. Compared with Case 1, the greenhouse gas emissions dropped by 5.9% every hour, which is 15,194.4 kg/h in Case 3. The total carbon emission is markedly decreased from 16,151.27 kg/h to 15,194.4 kg/h owing to the optimal allocation of ESSs. It should be noted that the emissions are 17,386 kg/h in Case 3 when the wind power reaches the maximum. The simulation results indicated that the operating range of the diesel engine should be taken into account when optimizing the configuration of the ESS.

Table 3. Carbon emission parameters of diesel engines.

Generator	d	e	f
Gen 1	22.983	−0.9	0.0126
Gen 2	25.505	−0.01	0.027
Gen 3	24.900	−0.005	0.0291
Gen 4	24.700	−0.004	0.0290
Gen 5	25.300	−0.0055	0.0271

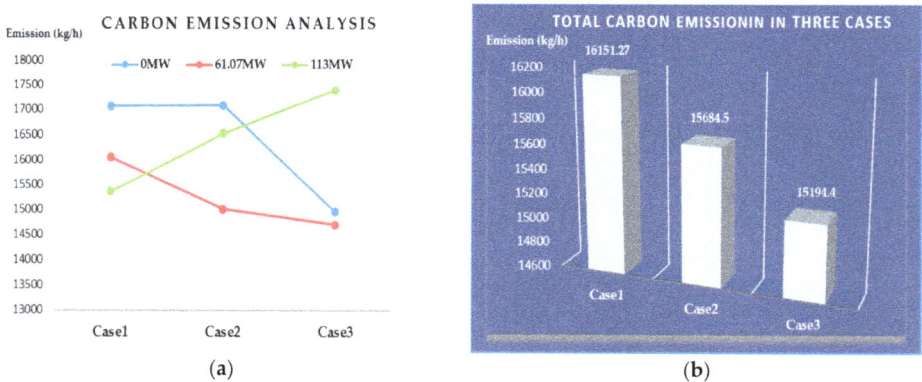

Figure 8. (**a**) Carbon emission considering wind distribution; (**b**) Total carbon emission.

As seen from the above analysis, the operation cost and carbon emission are greatly reduced by the distributed ESS configuration. Meanwhile, the real power loss of the system declines sharply from 26.4 MW to 10.7 MW. Eventually, Buses 12, 25, and 26 are found to be the best places to install ESSs with capacities of 30.7 MW, 18 MW, and 39.46 MW, respectively, which are demonstrated in Table 4.

Table 4. Optimal allocation of energy storage systems (ESSs).

ESS	BUS 12	BUS 25	BUS 26
Rated power (MW)	30.70	18.00	39.46
Capacity (MWh)	23.46	6.64	24.30

5. Conclusions

With the rapid development of renewable energy, it has become important to forecast electric energy when optimally planning a stable and economic power system. In this paper, a two-stage hybrid MOPSO that integrates with a back-propagation neural network is proposed to optimize the allocation of ESSs in order to reduce total cost and emissions. Furthermore, wind power distribution is discretized by a three-point estimation method, and the *P-V* curve is explored to select candidate buses for the installation of the ESSs. The simulation results show that (i) the dynamic prediction method is more suitable for forecasting wind power and load demand, which has the lowest number of errors; (ii) with the help of *P-V* sensitivity analysis, the proposed HMOPSO is able to search for the best placement and size for ESSs at a fast speed as well as minimize the total operation cost and improve voltage profiles; (iii) different from conventional analysis, the distributed ESSs can achieve minimum costs and greenhouse gas emissions for a wind power integrated system. In a future study, the air density and ambient temperature will be taken into account to improve the forecasting method.

Acknowledgments: This work was supported by the Natural Science Foundation of China under Grant No. 61174047 and by Fundamental Research Funds for the Central Universities under Grant No. HEUCFZ1305.

Author Contributions: All authors contributed to this collaborative work. Hai Lan and He Yin performed the research and discussed the results. Shuli Wen designed the optimal algorithm. Ying-Yi Hong, David C. Yu, and Lijun Zhang suggested the research idea and contributed to the writing and revision of the paper. All authors approved the manuscript.

Conflicts of Interest: The authors declare no conflict of interest.

Nomenclature

Acronyms

ANN	Artificial Neural Network
BFGS	Quasi-Newton
BPNN	Back-propagation Neural Network
ESS	Energy Storage System
HMOPSO	Hybrid Multi-objective PSO
LM	Levenberg Marquard
MOPSO	Multi-objective PSO
NAR	Non-linear Auto-regressive
NARX	Non-linear Auto-regressive model with exogenous inputs
NARX-BPNN	Non-linear Auto-regressive model with exogenous inputs—back-propagation neural network
O&M	Operation and Management
PEM	Point Estimation Method
PSO	Particle Swarm Optimization
PV	Photovoltaic
SOC	State of Charge
WT	Wind Turbine

Variables

a, b, c, c^w, c^e	Cost coefficients of different generations
C_E	Capacity of the ESS (MWh)
$C(P_d), C_w, C_e$	Cost of the diesel generator (\$/h), the WT (\$/h), and the ESS (\$/h)
$Cost_i$	Total operation cost at the i scenario (\$/h)
d, e, f	Carbon emission coefficients of diesel generations
E_E^{max}	Maximum charge/discharge demand of battery (MWh)
$Emission_i$	Carbon dioxide emissions at the i scenario (kg/h)
F_i	Total voltage fluctuations at the i scenario (V)
n	Total number of bus node
N	Number of diesel generators
n_c, n_d	Maximum charge/discharge rate of the ESS (C)
P_A	Output power of A (MW) $A \in \{d$(diesel generators), w(WT), e(ESS), l(load)$\}$
P_{dev}^A	Deviation from the schedule of A (MW)
P_F^B	Forecasted power of B (MW) $B \in \{w$(WT), l(load)$\}$
P_s	Hour-ahead scheduled power of diesel generators (MW)
$Prob_i$	Probability of target value at the i scenario
Q_{Gi}	Reactive power of diesel generator i
T_i	Tap of transformer i
u_{min}	Minimum battery charge/discharge rate (C)
V_k	RMS value of bus k voltage (V)
x	Wind speed (m/s)
Y	WT power in PEM model (MW)

References

1. Perez, M.J.R.; Fthenakis, V.M. On the spatial decorrelation of stochastic solar resource variability at long timescales. *Sol. Energy* **2015**, *117*, 46–58. [CrossRef]
2. Ogunjuyigbe, A.S.O.; Ayodele, T.R.; Akinola, O.A. Optimal allocation and sizing of PV/Wind/Split-diesel/Battery hybrid energy system for minimizing life cycle cost, carbon emission and dump energy of remote residential building. *Appl. Energy* **2016**, *171*, 153–171. [CrossRef]
3. Chen, C.; Duan, S.; Cai, T. Optimal allocation and economic analysis of energy storage system in microgrids. *IEEE Trans. Power Electron.* **2011**, *26*, 2762–2773. [CrossRef]
4. ElNozahy, M.S.; Tarek, K.; Abdel-Galil; Salama, M.M.A. Probabilistic ESS sizing and scheduling for improved integration of PHEVs and PV systems in residential distribution systems. *Electr. Power Syst. Res.* **2015**, *125*, 55–66. [CrossRef]
5. Bludszuweit, H.; Dominguez, N.J.A. A probabilistic method for energy storage sizing based on wind power forecast uncertainty. *IEEE Trans. Power Syst.* **2011**, *26*, 1651–1658. [CrossRef]
6. Wang, W.; Mao, C.; Lu, J. An energy storage system sizing method for wind power integration. *Energies* **2013**, *6*, 3392–3404. [CrossRef]
7. Wang, B.; Yang, Z.P.; Lin, F. An improved genetic algorithm for optimal stationary energy storage system locating and sizing. *Energies* **2014**, *7*, 6434–6458. [CrossRef]
8. Xiao, J.; Liang, H.; Yu, J.; Zhang, P.; Wang, X.; Yuan, S. A capacity optimization method for hybrid energy storage system considering SOC and efficiency. In Proceedings of the Renewable Power Generation Conference (RPG 2013), 2nd IET, Beijing, China, 9–11 September 2013; pp. 1–4.
9. Motaleb, A.M.A.E.; Bekdache, S.K.; Barrios, L.A. Optimal sizing for a hybrid power system with wind/energy storage based in stochastic environment. *Renew. Sustain. Energy Rev.* **2016**, *59*, 1149–1158. [CrossRef]
10. Antonanzas, J.; Osorio, N.; Escobar, R.; Urraca, R.; Martinez-de-Pison, F.J.; Antonanzas-Torres, F. Review of photovoltaic power forecasting. *Sol. Energy* **2016**, *136*, 78–111. [CrossRef]
11. Yan, J.; Liu, Y.; Han, S.; Wang, Y.; Feng, S. Reviews on uncertainty analysis of wind power forecasting. *Renew. Sustain. Energy Rev.* **2015**, *52*, 1322–1330. [CrossRef]
12. Hong, T.; Fan, S. Probabilistic electric load forecasting: A tutorial review. *Int. J. Forecast.* **2016**, *32*, 914–938. [CrossRef]

13. Claudio, M.; Ignacio, J.R.R.; Alfredo, F.J.L. Short-term forecasting model for aggregated regional hydropower generation. *Energy Convers. Manag.* **2014**, *88*, 231–238.
14. Okumus, I.; Dinler, A. Current status of wind energy forecasting and a hybrid method for hourly predictions. *Energy Convers. Manag.* **2016**, *123*, 362–371. [CrossRef]
15. Mazorra, A.L.; Pereira, B.; Lauret, P. Combining solar irradiance measurements, satellite-derived data and a numerical weather prediction model to improve intra-day solar forecasting. *Renew. Energy* **2016**, *97*, 599–610.
16. Fernando, A.O.P.; Jesús, F.B.; Juan, F.G.F. Failure mode prediction and energy forecasting of PV plants to assist dynamic maintenance tasks by ANN based models. *Renew. Energy* **2015**, *81*, 227–238.
17. Giorgi, M.G.D.; Congedo, P.M.; Malvoni, M. Photovoltaic power forecasting using statistical methods: Impact of weather data. *Meas. Technol.* **2014**, *8*, 90–97. [CrossRef]
18. Alonso-Montesinos, J.; Batlles, F.J.; Portillo, C. Solar irradiance forecasting at one-minute intervals for different sky conditions using sky camera images. *Energy Convers. Manag.* **2015**, *105*, 1166–1177. [CrossRef]
19. Bacher, P.; Madsen, H.; Nielsen, H.A. Online short-term solar power forecasting. *Sol. Energy* **2009**, *83*, 1772–1783. [CrossRef]
20. Zidar, M.; Georgilakis, P.S.; Hatziargyriou, N.D. Review of energy storage allocation in power distribution networks: Applications, methods and future research. *IET Gener. Transm. Distrib.* **2016**, *10*, 645–652. [CrossRef]
21. Suchitra, D.; Jegatheesan, R.; Deepika, T.J. Optimal design of hybrid power generation system and its integration in the distribution network. *Int. J. Electr. Power Energy Syst.* **2016**, *82*, 136–149. [CrossRef]
22. Ganguly, S. Multi-objective planning for reactive power compensation of radial distribution networks with unified power quality conditioner allocation using particle swarm optimization. *IEEE Trans. Power Syst.* **2014**, *29*, 1801–1810. [CrossRef]
23. Ramadan, H.S.; Bendary, A.F.; Nagy, S. Particle swarm optimization algorithm for capacitor allocation problem in distribution systems with wind turbine generators. *Int. J. Electr. Power Energy Syst.* **2017**, *84*, 143–152. [CrossRef]
24. Muhammad, K.; Abdollah, A.; Andrey, V.S.; Vassilios, G.A. Minimizing the energy cost for microgrids integrated with renewable energy resources and conventional generation using controlled battery energy storage. *Renew. Energy* **2016**, *97*, 646–655.
25. Du, Y. Optimal allocation of energy storage system in distribution systems. *Procedia Eng.* **2011**, *15*, 346–351.
26. Lan, H.; Dai, J.; Wen, S. Optimal tilt angle of photovoltaic arrays and economic allocation of energy storage system on large oil tanker ship. *Energies* **2015**, *8*, 11515–11530. [CrossRef]
27. Fichaux, N.; Poglio, T.; Ranchin, T. Mapping offshore wind resources: Synergetic potential of SAR and scatterometer data. *IEEE J. Ocean. Eng.* **2005**, *30*, 516–525. [CrossRef]
28. Wang, S.; Zhang, N.; Wu, L.; Wang, Y. Wind speed forecasting based on the hybrid ensemble empirical mode decomposition and GA-BP neural network method. *Renew. Energy* **2016**, *94*, 629–636. [CrossRef]
29. Haddad, S.; Benghanem, M.; Mellit, A.; Daffallah, K.O. ANNs-based modeling and prediction of hourly flow rate of a photovoltaic water pumping system: Experimental validation. *Renew. Sustain. Energy Rev.* **2015**, *43*, 635–643. [CrossRef]
30. Hussain, S.; Al-Alili, A. A new approach for model validation in solar radiation using wavelet, phase and frequency coherence analysis. *Appl. Energy* **2016**, *164*, 639–649. [CrossRef]
31. Saleh, F.; Hossein, R.; Alireza, H.; Hossein, I. A novel PSO (Particle Swarm Optimization)-based approach for optimal schedule of refrigerators using experimental models. *Energy* **2016**, *107*, 707–715.
32. Kennedy, J.; Eberhart, R. Particle swarm optimization. *IEEE Int. Conf. Neural Netw.* **1995**, *4*, 1942–1948.
33. Eberhart, R.; Kennedy, J. A new optimizer using particle swarm theory. In Proceedings of the International Symposium on MICRO Machine and Human Science, Ann Arbor, MI, USA, 4–6 October 1995; pp. 39–43.
34. Robert, J.H.; Nacer, K.M.; Aziz, N. Optimization of hybrid renewable energy systems (HRES) using PSO for cost reduction. *Energy Procedia* **2013**, *42*, 318–327.
35. Clerc, M.; Kennedy, J. The particle swarm-explosion, stability, and convergence in a multidimensional complex space. *IEEE Trans. Evol. Comput.* **2002**, *6*, 58–73. [CrossRef]
36. Deb, K.; Kalyanmoy, D. *Multi-Objective Optimization Using Evolutionary Algorithms*; John Wiley & Sons: Hoboken, NJ, USA, 2001.
37. Mohn, F.W.; Souza, A.C.Z. Tracing PV and QV curves with the help of a CRIC continuation method. *IEEE Trans. Power Syst.* **2006**, *21*, 1115–1122. [CrossRef]

38. 30 Bus Power Flow Test Case. Available online: http://www.ee.washington.edu/research/pstca/pf30/pg_tca30bus.htm (accessed on 1 August 1993).
39. Krishnamurthy, S.; Tzoneva, R. Investigation on the impact of the penalty factors over solution of the dispatch optimization problem. In Proceedings of the 2013 IEEE International Conference on Industrial Technology (ICIT), Cape Town, South Africa, 25–28 February 2013; pp. 851–860.

applied
sciences

MDPI

Article

Development of a Simulation Framework for Analyzing Security of Supply in Integrated Gas and Electric Power Systems

Kwabena Addo Pambour [1], Burcin Cakir Erdener [2], Ricardo Bolado-Lavin [3,*] and Gerard P. J. Dijkema [1]

[1] Energy and Sustainability Research Institute Groningen (ESRIG), Faculty of Science and Engineering, University of Groningen, Nijenborgh 6, 9747 AG Groningen, The Netherlands; kwabena.pambour@gmail.com (K.A.P.); g.p.j.dijkema@rug.nl (G.P.J.D.)

[2] Directorate (C) for Energy, Transport and Climate, Joint Research Centre, European Commission, Via Enrico Fermi 2749, I-21027 Ispra (VA), Italy; burcin.cakir@jrc.ec.europa.eu

[3] Directorate (C) for Energy, Transport and Climate, Joint Research Centre, European Commission, Westerduinweg 3, 1755 ZG Petten, The Netherlands

* Correspondence: ricardo.bolado-lavin@ec.europa.eu; Tel.: +31-224-565-349

Academic Editors: Josep M. Guerrero and Amjad Anvari-Moghaddam
Received: 22 November 2016; Accepted: 23 December 2016; Published: 5 January 2017

Abstract: Gas and power networks are tightly coupled and interact with each other due to physically interconnected facilities. In an integrated gas and power network, a contingency observed in one system may cause iterative cascading failures, resulting in network wide disruptions. Therefore, understanding the impacts of the interactions in both systems is crucial for governments, system operators, regulators and operational planners, particularly, to ensure security of supply for the overall energy system. Although simulation has been widely used in the assessment of gas systems as well as power systems, there is a significant gap in simulation models that are able to address the coupling of both systems. In this paper, a simulation framework that models and simulates the gas and power network in an integrated manner is proposed. The framework consists of a transient model for the gas system and a steady state model for the power system based on AC-Optimal Power Flow. The gas and power system model are coupled through an interface which uses the coupling equations to establish the data exchange and coordination between the individual models. The bidirectional interlink between both systems considered in this studies are the fuel gas offtake of gas fired power plants for power generation and the power supply to liquefied natural gas (LNG) terminals and electric drivers installed in gas compressor stations and underground gas storage facilities. The simulation framework is implemented into an innovative simulation tool named SAInt (Scenario Analysis Interface for Energy Systems) and the capabilities of the tool are demonstrated by performing a contingency analysis for a real world example. Results indicate how a disruption triggered in one system propagates to the other system and affects the operation of critical facilities. In addition, the studies show the importance of using transient gas models for security of supply studies instead of successions of steady state models, where the time evolution of the line pack is not captured correctly.

Keywords: combined simulation; power and gas interdependence; security of supply; transient gas simulation; scenario analysis; power system contingency

1. Introduction

Large scale energy infrastructures for natural gas and power play a crucial role for any well-functioning society. These infrastructures are systematically analyzed and controlled in order to understand their operational characteristics and to provide an energy efficient operation and a sufficient level of security of supply. However, ensuring the required level of security of supply is becoming more challenging, especially because of the increasing interconnections among the facilities in both systems.

The dependence of power generation on natural gas has increased the vulnerability of electric power systems to interruptions in gas supply, transmission, and distribution. Since the storage of gas on-site is not an option, as it is for coal and fuel oil, the direct gas delivery through pipelines becomes more critical during unexpected events in electricity systems like peak periods or disruptions. Particularly, short term problems caused by pipeline constraints that cause an inability of a generator to receive fuel gas can seriously affect security of power supply [1].

Another issue is the lack of predictability of renewable generation, which might increase the magnitude of imbalances in the gas system. Although the increasing share of renewables will cause a reduction of the power system dependency on natural gas, forecasting the amount of gas needed to serve Gas Fired Power Plants (GFPPs) will become more challenging due to growing penetration of variable resources. Additionally, shale gas production already had a significant impact on the deployment of new infrastructures, especially in the USA, where the installed capacity of GFPPs has increased enormously during the last years and is expected to continue increasing in the coming years [2]. This increase has obviously tightened the dependency of the electricity system on the gas system. This could also be the case in other regions of the world, including Europe, especially under scenarios of abundant shale gas and low carbon policies.

Not only is the power system dependent on gas, but also the gas system is dependent on power. A gas network consists of different facilities that depend on electrical power in order to maintain normal operation (e.g., electric driven compressors, liquefied natural gas (LNG) facilities, underground gas storage facilities, valves, regulators, gas meters). The usage of electric drivers in gas facilities is increasing due to advantages regarding environmental impacts and flexibility compared to gas turbines [3]. Moreover, increased availability, better control, improved energy efficiency, and shorter delivery times are other important and attractive advantages of electric drivers. Since the proper functioning of electric drivers requires a reliable power supply, gas system dependency on the power system can be considered critical. Additionally, the present advancement in the Power-to-Gas (P2G) technology, where excess power generation from renewable sources is used to produce hydrogen or synthetic natural gas (SNG) will significantly contribute to the coupling of both systems [4], since the power system will depend on the gas system as an energy storage provider.

Summarizing these aspects, it appears that interconnections between gas and electricity systems make the entire energy system vulnerable, since a disruption occurring in one system (e.g., an unexpected failure) may propagate to the other system and may possibly feed back to the system, where the disruption started. Tight relations are increasing the potential risk for catastrophic events, triggered by either intentional or unintentional disruptions of gas or electricity supply and possibly magnified by cascading effects. Analyzing both systems in an integrated manner and developing a combined assessment methodology is needed in order to know whether and how such interdependencies may contribute to the occurrence of large outages and to ensure the proper functioning of the energy supply system.

In this paper, we propose a simulation framework for assessing the interdependency of integrated gas and power systems in terms of security of supply. The framework combines a steady state AC-flow model with a transient hydraulic gas model and captures the physics of both systems. The data exchange between both models is established through a developed software application named SAInt (Scenario Analysis Interface for Energy Systems), which contains a graphical user interface for creating the network models and scenarios and for evaluating the simulation results. The proposed framework implemented in SAInt, is intended to be used by system operators, researchers, operational planners interested in analyzing the operation and interdependency of gas and electricity systems in terms of security of energy supply; i.e., to analyze the cascading impacts of contingencies on the operation of integrated gas and power systems and to assess system flexibilities by providing information on system abilities to react to changes.

To achieve these goals, the paper follows the following pattern. In Section 2, we give an overview of available models in the scientific literature addressing the analysis of combined gas and power systems and highlight the gaps in the literature we intend to fill. Section 3 discusses the different modeling aspects to be considered in a combined gas and power system model for assessing security of supply, while Section 4 elaborates the developed simulation framework and its implementation into a software application. In Section 5, we apply the proposed model to perform a contingency analysis on a real life sized test network. Finally, the conclusions are given in Section 6.

2. State of the Art

The area of analyzing interdependencies between gas and power systems is relatively new. It is encouraging that the number of publications on integrated gas and electricity systems found in the literature is increasing, although still limited. Comprehensive reviews of past publications can be found in [5–8]. The different types of analysis undertaken in integrated gas and power systems literature can be categorized as; economic and market perspective analysis, operation planning and control (e.g., optimization, demand response), design and expansion planning, and security analysis.

Studies on the medium and long-term economic evaluations aiming at exploring the interactions between the mechanisms of pricing of each carrier are reported in [7–17], where the influence of technical constraints is often ignored or taken into account in a simplified way. Additionally in [18], the authors proposed a dynamic model representation of coupled natural gas and electricity network markets to test the potential interaction with respect to investments while considering network constraints of both markets. In [19], two methodologies for coupling interdependent gas and power market models are proposed in a medium-term scope, where the two systems are formulated separately as optimization problems and the obtained primal dual information is utilized.

From the operational viewpoint, unit commitment models relating to short term security constrained operation of combined gas and power systems are developed in [20–22]. In [21], the authors considered the natural gas network constraints in the optimal solution of security constrained unit commitment (SCUC). Additionally dual fuel units are modelled for analyzing different fuel availability scenarios. In [22], the model proposed in [21] is extended using a quadratic function of pressure for describing the gas flow in pipelines and also including the gas consumption of the compressors. In [23], an economic dispatch model (ED) is developed for integrated gas and power systems. The security constraints for both systems are integrated in the ED which aims to minimize power system operating costs. The optimal power flow (OPF) of the coupled gas and power systems are investigated in [24–29]. A method for OPF and scheduling of combined electricity and natural gas systems with a transient model for natural gas flow is investigated in [27] and the solutions for steady-state and transient models of the gas system are compared. A multi-time period OPF model was developed for the combined GB electricity and gas networks in [28,29].

The impact of uncertainties on integrated gas and power system operation caused by variable wind energy is discussed in [30–33]. In [30] the impacts of abrupt changes of power output from GFPPS, to compensate variable power output from wind farms, on the Great Britain (GB) gas network is analyzed. In [32], the authors developed partial differential equation (PDE) model of gas pipelines to analyze the effects of intermittent wind generation on the fluctuations of pressure in GFPPs and pipelines. The coordination between the gas and power systems based on an integrated stochastic model for firming the variability of wind energy is presented in [33]. Gas transmission system constraints and the variability of wind energy is considered in the optimal short-term operation of stochastic power systems with a scenario based approach.

Studies considering the implementation of demand side response in order to mitigate the pressure of peak demand can be found in [34–37]. An operating strategy for short-term scheduling of integrated gas and power system is proposed in [36] while considering demand response and wind uncertainty.

In [37], the impact of demand side response on integrated gas and power supply systems in GB is analyzed for the time horizon from 2010 to 2050.

The problem of the design and expansion planning is addressed in [38,39] for the integrated gas and power systems at the distribution level and the transmission level, respectively.

Recently P2G has gained significant interest. A number of studies [4,40] have investigated the interdependencies introduced by P2G units on the integrated gas and power system operation in GB. The application of P2G for seasonal storage in gas networks was investigated in [41].

The security perspective including the reliability and the adequacy of integrated gas and power systems has gained significant interest due to increasing dependencies among the systems. Such studies may include the cascading effects of contingencies where the performance of the networks is reduced [8,42–45]. In [8], an integrated simulation model that aims at reflecting the dynamics of the systems in case of disruptions is proposed. While developing the integrated model, first gas and power systems are modeled separately and then linked with an interface.

Despite the growing interest in analyzing the integrated gas and power system in reliability aspects most of the studies on that area have used steady-state or successions of steady state formulations (i.e., supply and demand are assumed balanced at all time) to define each system in order to reduce the complexity of the problem [21,46,47]. However, these formulations could not reflect the different behavior of the two systems appropriately, since gas and power system dynamics evolve on very different timescales. Gas systems react slower to changes in the system, because of the larger system inertia, due to the quantity of gas accumulated in the pipelines, also referred to as linepack. Since steady state gas models do not account for the changes in linepack, these models are inadequate for describing the dynamic behavior of gas systems, when boundary conditions change over time (demand, supply, etc.) [48,49]. Capturing the dynamic behavior of gas systems correctly requires the use of transient models. Nevertheless, few references can be found in literature considering the integration of gas dynamics with electricity systems [27,49]. In [27], gas and electricity systems have been modeled in a coupled manner to assess the coordinated daily scheduling of interdependent gas and electricity transmission systems that are based on slow transient process of gas flow. However, the authors did not take into account the ability of GFPPs to change their output within the day. Moreover, the flexibility of the gas system to adapt itself to changing demands of GFPPs is not analyzed in the study. In [49] an integrated gas and electric flexibility model has been developed where a relevant flexibility metric is introduced to assess the ability of the gas transmission networks to react to changes in the power system, particularly, due to intermittent renewables. The proposed model uses both steady-state and transient gas analysis and electrical DC optimal power flow, where the bus voltages and reactive power balance are neglected. The simplification used in DC power model may provide too optimistic results, mainly because voltage profile of buses and reactive power has significant impacts on the system conditions when perturbed by failure events [8].

This study extends previous work in the area in several ways. First, to the best of our knowledge it is the first scenario-based integrated simulation tool to analyze the cascading effects of the contingencies for integrated gas and power systems in such detail. The proposed framework (referred to as SAInt) couples a transient gas hydraulic model, which considers sub-models of the most important facilities, such as compressor stations, LNG terminals and UGS facilities, with a steady state power model based on AC flow, where the transmission capacity, active-reactive generation and upper-lower limits on voltage magnitude are considered. The gas model is designed with a dynamic time step adaptation method which adapts the simulation time step in relation to the control mode changes in order to capture these changes with a higher time resolution. Moreover bidirectional interdependencies are modeled by considering the gas dependency of GFPPs and the power dependency of electric driven compressor stations, LNG terminals and UGS facilities. The proposed model focuses on integrated analysis of gas and electricity systems to achieve a sustainable energy system and to improve energy security, as well as aiming at developing a methodology to identify and assess the impact of interactions between gas and electric systems in terms of energy security.

3. Security of Supply in Integrated Gas and Power Systems

The interactions between gas and electric systems make it increasingly difficult to separate security of gas supply from security of electricity supply. The changes in the overall system due to all type of incidents affect the dynamic behavior and vulnerability of the integrated gas/electricity system. The degree of integrated power and gas system vulnerability depends on some external conditions like the level of power system dependency on GFPPs, power generation mixture of the region, weather conditions, natural disaster probabilities of the region, and failure probability of facilities in either of the systems, among other factors.

Generally speaking, large disruptions in gas systems affecting both power and non-power consumers are not so common. The gas system is well known as reliable and safe. However, there could be incidents resulting in curtailment of gas in some conditions which can immediately cause problems in the power system such as, unexpected increase in demand, freezing of wellheads and disruption of pipelines among others. In such cases, the delivery pressure needed by the facilities has to be taken into account. This is particularly important in recently deployed GFPPs using modern combustion turbines, which need higher gas pressure to operate compared to conventional combustion turbines. It should be noted that, even if the gas system had enough capacity to deliver gas to GFPPs at peak demand, the coincidence of peak demand for GFPPs and for conventional use (household, commercial, industrial) may result in a significant diminished pressure in pipelines, which eventually may produce interruptions in the electricity generation because of insufficient pressure.

In case of lack of gas supply in a GFPP, the possible solutions that may help bridge the gap of gas availability could be dual fuel capabilities or/and a variety of storage options (line-pack and UGS facilities close to consumption areas). However, the costs and feasibility of storage and fuel switching has to be analyzed in detail since sometimes they cannot be used as a solution in practice. In fact, quite frequently because of the cost of fuel-oil storage a dual fuel GFPP cannot switch to the alternative fuel due to lack of fuel stored on-site.

When the consequences and cascading effects of a disruption originating in one system and propagating to the other system are compared, the gas system is more resilient to local and short-term disruptions compared to the electricity system. The main reason for this is that, in addition to the existence of the linepack as short-term storage, the majority of compressor stations are still powered by gas turbines, which keeps the pressure profile within limits, allowing continued operation. Furthermore, in case electric driven compressors are installed, a back-up power system (typically diesel) is usually available to protect the system from power outages. A massive power failure would generally have no serious effect on the physical pipeline facilities, provided that it does not last too long. Compressor stations that utilize electric drivers would be the most affected and have to be analyzed carefully.

When analyzing and modelling integrated gas and electricity systems, there are several issues that have to be addressed mainly due to the differences in the structure of the systems. For instance, the failure response of the power and gas system infrastructures is significantly different. A technical failure in the power system infrastructure can result in an immediate loss of service from a generating unit or a transmission line, that can, under some extreme conditions, propagate loss of service to the electric customers due to cascading effects. On the contrary, most technical failures in gas systems (e.g., pipeline rupture, failure in compressor station or storage facility etc.) result in a locally or regionally reduced network capacity rather than an entire loss of service to the gas consumers [1]. This capacity reduction might result in curtailments of gas delivery to customers according to their priority level of service. Another important distinction is the different dynamic behavior of the two systems. Electricity travels almost instantaneously and cannot be stored economically in large quantities in current power systems, with the only exception of hydraulic pumping power stations, whose availability is very much limited in a significant number of countries. In case of disruptions, the response time of the power system is quite small and basically the transmission line flows satisfy the steady-state algebraic equations. On the contrary, the gas flow in pipelines is a much slower process, with gas velocities below 15 m/s, resulting in a longer response time in case of disruptions.

In particular, high-pressure transmission pipelines have much slower dynamics due to the large volumes of gas stored in the pipelines. This quantity of gas cannot be neglected when simulating the dynamics in a gas transmission system; in fact the line pack in the pipeline increases the flexibility of the gas system to react to short term fluctuations in demand and supply. This information is important especially in the modeling stage, since different timing of the systems needs to be considered during the simulation process.

Based on the information above, a simulation framework is proposed that allows simulating integrated gas and power systems in a realistic way, emphasizing the integration and communication between the networks. The architecture of the simulator is explained in detail in the next section.

4. Methodology

In this section, we elaborate the different models and methods used in the proposed simulation framework for analyzing the interdependence between gas and power systems. In the first part, we derive the physical equations describing the behavior of both systems independently. Next, we elaborate the coupling equations describing the most relevant interconnections between the two energy systems. Finally, we integrate the individual models together with the coupling equations into a single integrated simulation framework and describe the algorithm and the communication and synchronization between the simulators in the course of the solution process of the combined energy system.

4.1. Power System Model

A power transmission system is described by a directed graph $G = (V, E)$ consisting of a set of nodes V and a set of branches E, where each branch $e \in E$ represent a transmission line or a transformer and each node $i \in V$ a connection point between two or more electrical components, also referred to as bus. At some of the buses power is injected into the network, while at others power is consumed by system loads.

Transmission lines and transformers, can be described by a generic per-phase equivalent π-circuit model depicted in Figure 1, which reflects the basic properties of both components, such as resistance R_{ft}, reactance X_{ft}, line charging susceptance b_{ft}, transformer tap ratio t_{ft} and phase shift angle ϕ_{ft}.

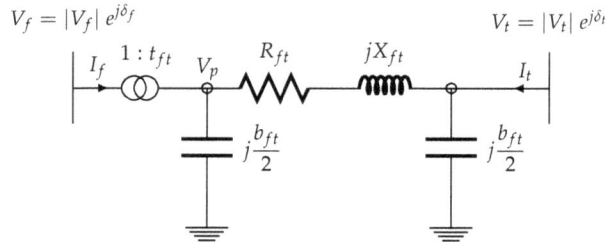

Figure 1. Generic branch model (π-circuit) for modeling transmission lines ($t_{ft} = 1$ & $\phi_{ft} = 0$), in-phase transformers ($\phi_{ft} = 0$) and phase-shifting transformers ($\phi_{ft} \neq 0$). The transformer tap ratio is modeled only on the from-Bus side of the branch model.

From the π-circuit model, we can derive for each branch $e \in E$ a branch admittance matrix $\mathbf{Y}_{br}$, which relates the complex from-bus and to-bus current injections I_f & I_t, respectively, to the complex from-bus and to-bus voltages V_f & V_t, respectively, as follows:

$$\begin{bmatrix} I_f \\ I_t \end{bmatrix} = \begin{bmatrix} a_{ft}^2 (y_{ft} + \frac{b_{ft}}{2}) & -t_{ft}^* \cdot y_{ft} \\ -t_{ft} \cdot y_{ft} & a_{tf}^2 (y_{ft} + \frac{b_{ft}}{2}) \end{bmatrix} \begin{bmatrix} V_f \\ V_t \end{bmatrix} \tag{1}$$

with

$$t_{ft} = \frac{V_p}{V_f} = a_{ft}\,e^{j\phi_{ft}}, \; a_{ft} = \frac{|V_p|}{|V_f|}, \; y_{ft} = \frac{1}{R_{ft} + jX_{ft}} = \frac{1}{Z_{ft}} \tag{2}$$

The elements of the branch admittance matrices can be used to assemble the bus admittance matrix $\mathbf{Y_{bus}}$ which describes the relation between the vector of complex bus current injections $\mathbf{I}$ to the vector of complex bus voltages $\mathbf{V}$ for the entire power network.

$$\mathbf{I} = \mathbf{Y_{bus}} \cdot \mathbf{V}, \quad \mathbf{Y_{bus}} = \left[Y_{ij}\right]^{N_b \times N_b} \tag{3}$$

The steady state power balance in the power system is derived from Kirchhoff's Current Law (KCL, i.e., all incoming and outgoing currents at a bus must sum up to zero) applied to each bus in the network, which yields the following complex power balance matrix equation for the entire network:

$$\mathbf{S} = \mathbf{V} \cdot \mathbf{I^*} \Rightarrow (\mathbf{P_G} - \mathbf{P_D}) + j(\mathbf{Q_G} - \mathbf{Q_D}) = \mathbf{V} \cdot \mathbf{Y_{bus}^*} \cdot \mathbf{V^*} \tag{4}$$

where the left hand side describes the active $\mathbf{P}$ and reactive $\mathbf{Q}$ power injections/extractions at generation/load buses, respectively, and the right hand side the incoming and outgoing apparent power flows from transmission lines and transformers.

The operation of a power system is restricted by a number of constraints imposed by technical components and stakeholders (producers, consumers, regulators etc.) involved in the power supply chain. Transmission lines, for instance, can only transport a limited amount of power due to thermal restrictions, while the operation of power plants is limited by the capability curves of the installed generators. The power transmission system operator (TSO) is responsible for respecting these constraints, while operating the system in an economic and secure manner. The real time power dispatch in an electric power system can be described by a steady state AC-optimal power flow model (AC-OPF) [50], which is expressed by the following non-linear inequality constrained optimization model:

$$\min_{\mathbf{X}} \quad f(\mathbf{X}) = \sum_{i=1}^{N_g} c_{0,i} + c_{1,i}\,P_{G,i} + c_{2,i}\,P_{G,i}^2 \tag{5}$$

$$s.\,t. \quad G_{P,i}(\mathbf{X}) = P_i(\mathbf{V}) - P_{G,i} + P_{D,i} = 0, \quad i = 1 \dots N_b \tag{6}$$

$$G_{Q,i}(\mathbf{X}) = Q_i(\mathbf{V}) - Q_{G,i} + Q_{D,i} = 0, \quad i = 1 \dots N_b \tag{7}$$

$$P_i(\mathbf{V}) = \sum_{j=1}^{N_b} |V_i||V_j||Y_{ij}|\cos(\delta_i - \delta_j - \theta_{ij}), \quad i = 1 \dots N_b \tag{8}$$

$$Q_i(\mathbf{V}) = \sum_{j=1}^{N_b} |V_i||V_j||Y_{ij}|\sin(\delta_i - \delta_j - \theta_{ij}), \quad i = 1 \dots N_b \tag{9}$$

$$H_k^f(\mathbf{X}) = S_k^{f*} \cdot S_k^f - S_k^{max2} \le 0, \quad k = 1 \dots N_l \tag{10}$$

$$H_k^t(\mathbf{X}) = S_k^{t*} \cdot S_k^t - S_k^{max2} \le 0, \quad k = 1 \dots N_l \tag{11}$$

$$\delta_i = \delta_i^{ref}, \quad i = i_{ref} \tag{12}$$

$$|V_i^{min}| \le |V_i| \le |V_i^{max}|, \quad i = 1 \dots N_b \tag{13}$$

$$P_{G,i}^{min} \le P_{G,i} \le P_{G,i}^{max}, \quad i = 1 \dots N_g \tag{14}$$

$$Q_{G,i}^{min} \le Q_{G,i} \le Q_{G,i}^{max}, \quad i = 1 \dots N_g \tag{15}$$

where the decision variables expressed by vector **X**

$$\mathbf{X} = \begin{bmatrix} \Delta & \mathbf{V_m} & \mathbf{P_G} & \mathbf{Q_G} \end{bmatrix}^T \tag{16}$$

are the set of bus voltage angles Δ, bus voltage magnitudes $\mathbf{V_m}$ and active and reactive power generation $\mathbf{P_G}$ and $\mathbf{Q_G}$, respectively. Equation (5) is a scalar quadratic objective function, which describes the total operating costs for each committed generation unit in terms of its active power generation, while the non-linear equality constraints expressed by Equations (6)–(9) describe the set of active and reactive power balance equations derived from matrix Equation (4). Equations (10) and (11) are non-linear inequality constraints, which describe the transmission capacity limits S_k^{max} for each line, while the upper and lower limits of the decision variables are described by Equations (13)–(15). For each isolated sub network one bus is chosen as the voltage angle reference (see Equation (12)), i.e., the voltage angle of the reference bus is set to zero.

The described AC-OPF model is implemented into the open source power flow library MATPOWER [51], which we utilize as the power system simulator in the context of the proposed simulation framework.

4.2. Gas System Model

Similar to the power system network, the gas network is described by a directed graph $G = (V, E)$ composed of nodes V and branches E. Facilities with an inlet, outlet and flow direction are modeled as branches, while connection points between these branches as well as entry and exit stations are represented by nodes. Branches, in turn, can be distinguished between active and passive branches. Active branches represent controlled facilities, which can change their state or control during simulation, such as compressor stations, regulator stations and valves, while passive branches, such as pipelines and resistors represent facilities or components whose state is fully described by the physical equations, derived from the conservation laws. A description of the different branch types is given in Table 1. Nodes can also be differentiated according to their function into supply, demand, storage and junctions. A description of the different node types and their corresponding node facilities is given in Table 2.

The gas model proposed in this study includes sub-models of all important facilities comprising a gas transport system, such as pipelines, compressor stations (CS), production fields (PRO), cross-border import (CBI) and export stations (CBE), city gate stations (CGS), stations of direct served customers (GFPPs, IND), liquefied natural gas (LNG) regasification terminals and underground gas storage (UGS) facilities. The model is able to capture appropiately [52,53] the reaction of gas transport systems to load variations (i.e., daily and seasonal changes of gas demands at offtake points) and disruption events (e.g., loss of supply from an entry point, failure in a compressor station, etc.) with moderate computational cost, taking into account the physical laws governing the dynamic behavior of gas transport systems. The accuracy of the proposed gas model has been confirmed in [52,53], where it is benchmarked against a commercial software package. In the following we give a brief description of the physical equations used fro describing the gas system. We refer to previous publications, for more details on the gas model implemented in SAInt [52–54].

The dynamic behavior of a gas transport system is predominantly determined by the gas flow in pipelines. The set of non-linear hyperbolic partial differential equations (PDE) describing the transient flow of natural gas in pipelines are derived from the law of conservation of mass, momentum and energy and the real gas law.

Table 1. Basic elements comprising gas transport networks.

Element Types	Description
Passive Elements	
pipe	models a section of a pipeline, basic properties are length, diameter, roughness and pipe efficiency
resistor	models passive devices that cause a local pressure drop (e.g., meters, inlet piping, coolers, heaters, scrubbers etc.)
Active Elements	
compressor	models a compressor station with generic constraints, allows the specification of a control mode of the station (e.g., outlet pressure control, inlet pressure control, flow rate control etc.)
regulator	models a pressure reduction and metering station located at the interface of two neighboring networks with different maximum operating pressures, allows the specification of a control mode of the station (e.g., outlet pressure control, inlet pressure control, flow rate control etc.)
valve	models a valve station, which is is either opened or closed

Table 2. Classification and characteristics of nodes in the network.

Node Type	Description	Facilities
demand $L > 0$	point, where gas is extracted from the network, connected facilities are typically flow or pressure controlled	CGS, CBE, GFPP, IND
supply $L < 0$	point, where gas is injected into the network; connected facilities are typically flow or pressure controlled; for LNG regasification terminals the working gas inventory is monitored and the flow rate is reduced in case of low inventory	PRO, CBI, LNG
storage $L \geq 0 \ or \ L \leq 0$	point, where gas is injected or extracted from the network and where the maximum supply/loads depend on the working gas inventory, which is monitored along the transient simulation; connected facilities are typically flow or pressure controlled	UGS
junction $L = 0$	point, where a topological change or a change in pipe properties occurs (e.g., diameter, inclination); no specific control	-

Applying these laws on an infinitesimal control volume (*CV*) of a general pipeline with a constant cross-sectional area A and an infinitesimal length dx (see Figure 2) and assuming the parameters describing the gas flow dynamics along the pipe coordinate x are averaged over A, yields the following set of fundamental Partial Differential Equations (PDEs) describing the gas flow through pipelines (the assumption of averaging the flow parameters over the cross-sectional area can be justified as long as the pipe length L is much greater than the pipe diameter D which is the case in transmission networks where $\frac{D}{L}$ is of order $O(10^{-5})$ or lower):

Law of Conservation of Mass—Continuity Equation:

$$\frac{\partial \rho}{\partial t} + \frac{\partial (\rho v)}{\partial x} = 0 \tag{17}$$

Newton's Second Law of Motion—Momentum Equation:

$$\underbrace{\frac{\partial(\rho v)}{\partial t}}_{\text{inertia}} + \underbrace{\frac{\partial(\rho v^2)}{\partial x}}_{\text{convective term}} + \underbrace{\frac{\partial p}{\partial x}}_{\text{pressure force}} + \underbrace{\frac{\lambda \rho v|v|}{2D}}_{\text{shear force}} + \underbrace{\rho g \sin\alpha}_{\text{force of gravity}} = 0 \tag{18}$$

First Law of Thermodynamics - Energy Equation:

$$\frac{\partial}{\partial t}\left[\left(c_v T + \frac{1}{2}v^2\right)\rho A\right] + \frac{\partial}{\partial x}\left[\left(c_v T + \frac{p}{\rho} + \frac{1}{2}v^2\right)\rho u A\right] + \rho u A g \sin\alpha = \dot{\Omega} \tag{19}$$

Real Gas Law - State Equation:

$$\frac{p}{\rho} = Z R T \tag{20}$$

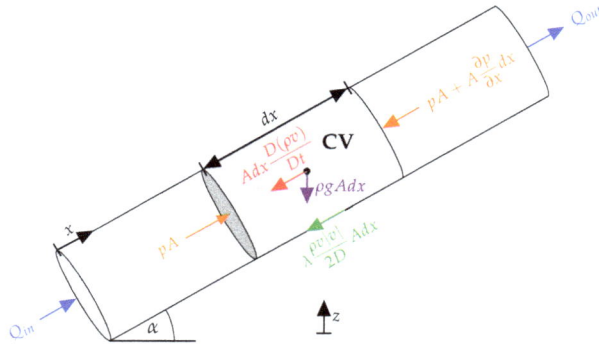

Figure 2. Forces acting on a control volume in a general gas pipeline.

The fundamental equations are typically simplified by adapting them to the prevailing conditions in transport pipelines. The most common assumptions are isothermal flow (i.e., constant temperature in time and space, thus, energy equation is redundant and can be neglected) and small flow velocities (i.e., relatively small Mach numbers, thus, convective term in momentum equation is negligible compared to the other terms), which applied to the above equations yields the following set of non-linear hyperbolic PDEs:

$$\frac{\partial p}{\partial t} = -\frac{\rho_n c^2}{A}\frac{\partial Q_n}{\partial x} \tag{21}$$

$$\frac{\partial p}{\partial x} = -\frac{\rho_n}{A}\frac{\partial Q_n}{\partial t} - \frac{\lambda \rho_n^2 c^2}{2DA^2 p}|Q_n|Q_n - \frac{g\sin\alpha}{c^2}p \tag{22}$$

with

$$c^2 = \frac{p}{\rho} = ZRT, \quad M = \rho v A = \rho_n Q_n$$

The above PDEs express the physical behavior of the gas flow in each pipe section in the gas model. We can integrate the set of PDEs for the entire network into one coupled equation system by applying the following integral form of the continuity equation to a nodal control volume V_i in the network, assuming all pipelines in the network are divided into a finite number of pipe segments:

$$\frac{V_i}{\rho_n c^2}\frac{dp_i}{dt} = \sum_{j=1}^{k} a_{ij}Q_{ij} - L_i, \quad i = 1\dots N_n \tag{23}$$

with

$$V_i = \frac{\pi}{8}\sum_{j=1}^{k} D_{ij}^2\,\Delta x_{ij}$$

Equation (23) can be expressed for each nodal control volume V_i in the network, resulting in N_n set of equations with $2N_n + M_b$ unknown state variables, where N_n and M_b denote the number of nodes and branches, respectively. Thus, $N_n + M_b$ additional independent equations are required in order to close the entire problem. These equations are provided by the pressure drop equation for each pipe segment derived in Equation (22) and the equations describing the control modes and active constraints of non-pipe facilities [52]. The differential equations can be discretized using the following fully implicit finite difference approximations for the state variables p, Q and L and their time and space derivatives:

$$\frac{\partial U}{\partial t} = \frac{U_i^{n+1} - U_i^n}{\Delta t}, \frac{\partial U}{\partial x} = \frac{U_{i+1}^{n+1} - U_i^{n+1}}{\Delta x}, U = \frac{U_{i+1}^{n+1} + U_i^{n+1}}{2} \tag{24}$$

The resulting (non-)linear finite difference equations and control equations of non-pipe facilities are solved iteratively by a sequential linearisation method [55]. For more details on the equations describing the control and active constraints of non-pipe facilities and the algorithm for solving the gas model we refer to [52–54].

Furthermore, we use the following expression for computing the quantity of gas stored in each pipe section (line pack):

$$LP(t) = \frac{A}{\rho_n \cdot c^2}\int_{x=0}^{x=\Delta x} p(x,t)\,dx = \frac{\Delta x\,p_m(t)\,A}{\rho_n \cdot c^2} \tag{25}$$

with

$$p_m(t) = \frac{2}{3}\frac{p_1(t)^2 + p_1(t)\cdot p_2(t) + p_2(t)^2}{p_1(t) + p_2(t)} \tag{26}$$

where p_m is the mean pressure in the pipe section and, p_1 and p_2 are the inlet and outlet gas pressure, respectively. The ramping of a GFPP depends on the availability of line pack in the hydraulic area (sub network bounded by controlled facilities) the GFPP is connected to. We consider the availability of line pack by setting a minimum nodal gas pressure threshold for the corresponding GFPP node. Since line pack is linearly correlated to the mean gas pipeline pressure, GFPPs operate only if a specific line pack level equivalent to the specified minimum pressure is available.

The presented gas model is implemented into the simulation tool SAInt, which we use as a simulator for the gas model in the proposed simulation framework.

4.3. Interconnection between Gas and Power Systems

As discussed in the previous sections, gas and electric power systems are physically interconnected at different facilities. In this paper, we consider the most important connections between both systems as follows:

1. *Power supply to electric drivers installed in gas compressor stations:*
    The electric power consumed by the compressor station can be described by the following

expression (derived from the first and second law of thermodynamics for an isentropic compression process) describing the required driver power $P_{D,i}^{CS}$ for compressing the gas flow Q from inlet pressure p_1 to outlet pressure p_2 [56,57]:

$$P_{D,i}^{CS} = f \frac{\kappa}{\kappa - 1} \frac{Z_1 T_1 R \rho_n Q}{\eta_{ad} \eta_m} \left[\frac{p_2}{p_1}^{\frac{\kappa - 1}{\kappa}} - 1 \right], \ i = 1 \ldots N_{CS} \tag{27}$$

where f is a factor describing the fraction of total driver power provided by electric drivers, η_{ad} the average adiabatic efficiency of the compressors, η_m the average mechanical efficiency of the installed drivers, p_2 the outlet pressure, p_1, Z_1, T_1 the inlet pressure, compressibility factor, temperature, respectively, R the gas constant, κ the isentropic exponent.

The power supply of the gas network is added to the active power demand in the electric model.

2. *Electric power supply to LNG terminals and UGS facilities:*
   We capture this interaction by assuming a generic linear function in terms of the regasification or withdrawal rate L_{rw}, respectively:

$$P_{D,i}^{rw} = k_{i,0} + k_{i,1} \cdot L_{rw,i} \tag{28}$$

3. *Fuel gas offtake from gas pipelines for power generation in GFPPs:*
   The required fuel gas $L_{GFPP,i}$ for active power generation $P_{G,i}$ at plant i can be expressed in terms of the thermal efficiency η_T of the GFPP and the gross calorific value GCV of the fuel gas, as follows:

$$L_{GFPP,i} = \frac{P_{G,i}}{\eta_T \cdot GCV}, \quad i = 1 \ldots N_{GFPP} \tag{29}$$

4.4. Integrated Simulation Framework for Security of Supply Analysis

The modeling framework carried out within SAInt considers the integrated gas and electricity transmission network under cascading outage contingency analysis. The cascading outages are investigated when the gas or electricity system has just experienced a disruption, like a shortage in supply or transmission capacity. The framework comprises of

(i)   a simulator (MATPOWER) for solving an AC-OPF for the power system,
(ii) a transient hydraulic gas simulator (SAInt) for the gas system which includes sub-models of all relevant pipe and non-pipe facilities
(iii) and an interface (SAInt) which handles the communication and data exchange between the two isolated simulators.

SAInt is composed of two separate modules, namely, SAInt-API (Application Programming Interface) and SAInt-GUI (Graphical User Interface). The API, is the main library of the software and contains the solvers, routines and classes for instantiating the different objects included in gas and electric systems. In order to perform power flow calculations and to extend the functionality of the software, the API has been linked to MATLAB using the Matlab COM Automation Server. This link has been used to establish a communication between the Matlab-based open source power flow library MATPOWER [51] and SAInt-API. This allows the execution of AC-Power Flow and AC-OPF with MATPOWER and the evaluation and visualization of the obtained results using SAInt-GUI [52]. For more information on SAInt we refer to previous publications [52,54].

The proposed simulation framework is illustrated in the flow diagram depicted in Figure 3, which is explained further below.

The power model proposed in this paper is designed to provide a realistic representation of the behavior of an actual power system when subjected to contingencies. Cascading effects of contingencies in the power grid are very complex phenomenona, and identifying the typical mechanisms of cascading failures and understanding how these mechanisms interact during blackouts is an important research area [58–63]. Potential mechanisms that might be modeled include overloaded line tripping by impedance relays due to the low voltage and high current operating conditions, line tripping due to loss of synchronism, the undesirable generator tripping events by overexcitation protection, generator tripping due to abnormal voltage and frequency system condition, and under-frequency or under-voltage load shedding. For each additional mechanism of cascading failure included in a model, assumptions must be made about how the system will react to these rarely observed operating conditions.

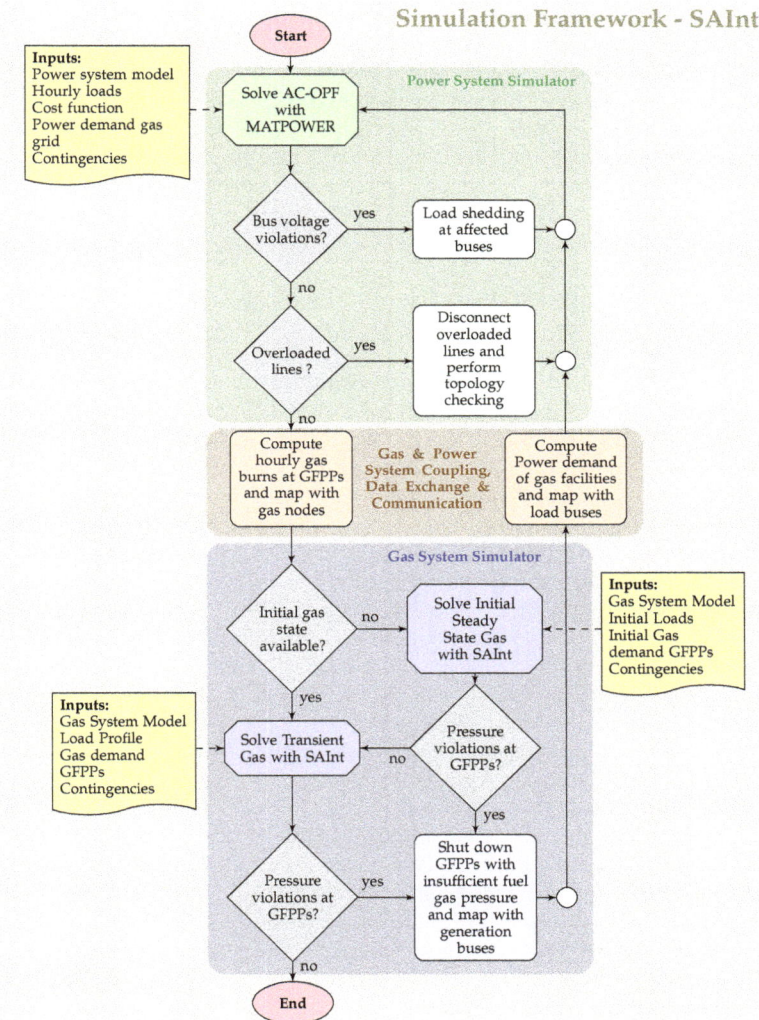

Figure 3. Flow chart of the proposed Simulation Framework SAInt, showing the implemented algorithm.

This paper introduces a steady state AC-flow model which is adapted to reflect a set of corrective actions performed by TSOs when trying to return the system to a stable operating condition after a contingency.

While the initial contingency can usually be considered as being a random event, an interaction of cascading failure mechanisms exists in the subsequent events. For example, the loss of critical components such as tripping of transmission lines creates load redistribution to other components, which might become overloaded. The overall network is then weakened due to the stress on remaining elements, possibly leading to an instability. If corrective action plans are not applied quickly further failures might be created as a consequence leading to a blackout. In this paper, this cascading failure phase, starting with the initiating event is modeled, where the cascading contingencies occurrence are affected by operator actions and the times between subsequent events are considered in a range of tens of seconds to 1 h. Various system adjustments that are considered include the post-contingency redispatch of active and reactive resources, cascade tripping of an overloaded transmission line, tripping or re-dispatching of generators due to load/generation imbalance, and load shedding at load buses to prevent a complete system blackout when insufficient voltage magnitudes are observed.

The initial state of the model is obtained by solving the standard AC optimal power-flow problem as described in Equations (5)–(15), which yields the optimum hourly generator dispatch for given hourly loads, cost functions for each generator and bus voltage and line loading constraints. To execute this task MATPOWER 6.01b AC-OPF algorithm is applied [51].

Any change from the initial state caused by a contingency event, such as a (simultaneous) failure of one or more transmission lines, failure of a generation unit or decreased amount of generation capacity due to lack of gas supply, can be introduced in the model by defining a scenario event for the corresponding facility, which is composed of an event time, an event parameter and its corresponding value.

Whenever a contingency is observed in the system, an imbalance between total generation and total load may occur. In order to re-balance the system, the model redistributes the missing or excess power to the remaining facilities in the power grid. The power re-dispatch is obtained by running the AC-OPF model, while considering the new topology triggered by a previous disruption (e.g., lines and generation units may be disconnected). However, since the system is under a stressed state, the AC-OPF algorithm may deliver an infeasible solution, that does not satisfy the convergence criteria, since system constraints such as line overloading or voltage limits cannot sustain the desired system loads. In order to allow the system to find a converged solution, the bus voltage ($|V| \geq |V^{min}|$) and line capacity constraints ($S_f \cdot S_f^* \leq S^{max2}$ & $S_t \cdot S_t^* \leq S^{max2}$) in the standard AC-OPF formulation are relaxed for the re-dispatching process. The re-dispatching process is followed by a two step feasibility checking procedure. In step one, bus voltage violations are mitigated by performing load shedding at the affected buses and recomputing the relaxed AC-OPF until no voltage violations are detected, so called under-voltage load shedding. The model assumes that there is enough time for the operator to implement under voltage load shedding to prevent a voltage collapse which is the root cause of most of the major power system disturbances [64–66]. The model sheds load in blocks of 2% for the corresponding bus until the relaxed bus voltage constraint is satisfied. If a violation is not eliminated although the load sheds more than 50% of its original load, we assume complete failure of the affected bus and set the load value to zero [8]. The second step of the feasibility checking procedure follows after all bus voltage violations have been remedied in the first step. During the re-dispatching process new failures may occur at certain components as they become overloaded. In this paper the overloads are aimed to be strictly avoided for all component contingencies. This means that it is assumed that the probability for line trip is 1 when line flow exceeds its thermal capacity with a tolerance parameter. The second step involves disconnecting overloaded transmission lines from the power grid and recomputing the relaxed AC-OPF until a feasible solution is obtained. It should be noted that, the connectivity of the network is checked in every simulation step prior to the AC-OPF computation in order to detect isolated facilities. The algorithm used for checking the connectivity is based on the well-known minimum spanning tree algorithm and is described in detail in [8].

After obtaining a feasible solution for the power system, the resulting hourly power generation of GFPPs is converted into a hourly gas demand profiles and provided as input to the gas model. The gas model needs an initial state for running the transient simulation. This state can either be a solution of a steady state simulation or the terminal state of a transient simulation. If an initial state is not available the algorithm uses the initial loads of the generated gas demand profiles for GFPPs to compute a steady state solution. This solution is then used as an initial state for the transient simulation. After each transient or steady state simulation the algorithm checks if the fuel gas pressures at GFPP nodes are sufficient to operate the facilities. If an insufficient fuel gas pressure is detected, the affected GFPP is shut-down and the power system model is recomputed. The algorithm is terminated if no pressure violations are detected after the transient gas simulation. Finally, the amount of energy not supplied is calculated as an indicator of the impact of the disruption event.

The gas and electric model described above are connected through an interface which enables the communication and data exchange between the two simulators (i.e., MATPOWER as power system simulator and SAInt as transient gas simulator, see Figure 3). The time integration of the combined model is performed separately for both systems and the interconnection between both systems is established through data exchange at discrete time and space points.

The timing of the power model is based on the discrete event simulation concept. It is assumed that the configuration of the power system (e.g., the state of generation units and lines) remains unchanged between events and changes only at the time of the specific event. If no events are scheduled or triggered in the course of the simulation the time step of the power system corresponds to a reference time step of 1 hour.

In contrast to the power system, the time integration of the transient gas model, is based on a dynamic time step adaptation method (DTA), which adapts the time resolution with respect to the control changes of controlled gas facilities during the solution process. The DTA allows capturing rapid changes in the gas system (shut-down of a power plant or compressor station etc.) with a higher time resolution. In this context, the gas model can be viewed as a quasi-continuous system, where the values of the state variables (i.e., nodal pressure p, element flows Q and nodal loads L) between two discrete time points are approximated by linear interpolation. If no events are scheduled or triggered in the course of the simulation the time step of the gas system corresponds to a reference time step of 15 min.

The gas and power system simulator used in the simulation framework have both been tested and verified. The gas simulator SAInt was benchmarked against a commercial software in previous publications [52,53] and the power system simulator MATPOWER [51] is well known and accepted by the scientific community.

In the following section, the proposed framework is applied to perform a contingency analysis for an integrated gas and power system network.

5. Model Application

In this section, an integrated gas and power network is constructed to demonstrate the previously discussed simulation framework implemented in SAInt. Three supply side scenarios (one non-disrupted scenario (base case) and two supply disruption scenarios) are presented in order to demonstrate the value of the proposed framework and to stress the importance of modeling the interdependence between gas and power systems with respect to security of supply.

The proposed scenarios are performed on the test network depicted in Figure 4. (The test network applied in this paper is a model of a real gas and electric power network of an European region. Due to confidentiality reasons and the sensitivity of the presented results, the topology and facility names of the real network have been disguised. The network topology and properties used for the computations, however, are realistic data for the combined network). The scenarios are composed of a number of extreme events causing more than two network facilities to be deactivated or to cascade out of service.

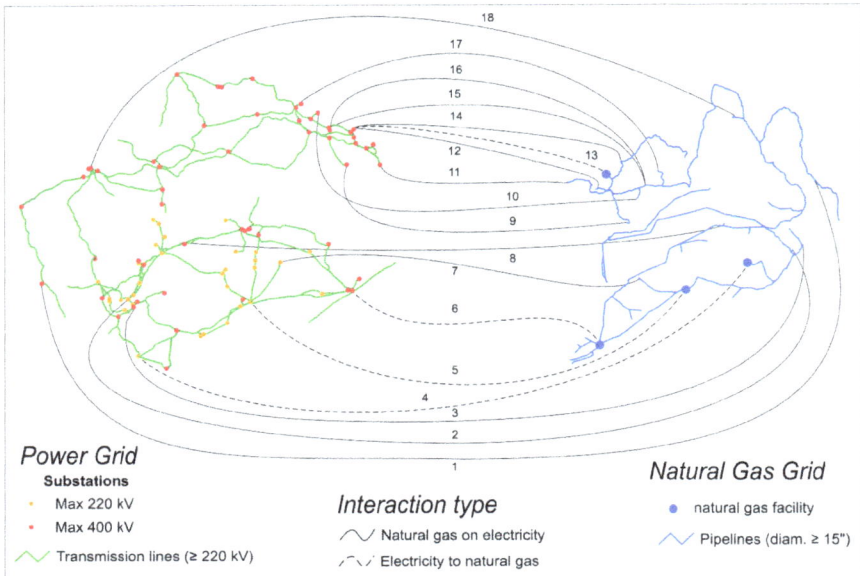

Figure 4. Integrated gas and power network applied in the case study. Map shows a real network of an European region, which has been disguised due to confidentiality reasons. The network data and properties used for the case studies, however, are original input data for the actual network. The solid black lines (lines 1–3, 7–12, 14–18) represent interconnections between Gas Fired Power Plants (GFPPs) in the power grid (**left**) and their fuel gas offtake points in the gas grid (**right**), while the dashed black lines (4–6, 13) represent interconnections between electric buses in the power grid (**left**) supplying electric power to connected facilities in the gas grid.

The sample network includes a power grid with 158 buses, 62 generating units with 22,076 (MW) installed capacity based on different generation mix that mainly consists of lignite (33%), natural gas (28%), coal (20%), wind power (7%) and others (12%). The transmission system consists of 194 high voltage transmission lines with total line length of approx. 8000 (km). The base voltage levels for the transmission lines are distinguished between 200 (kV) and 400 (kV).

The solution of the AC-OPF equations requires the knowledge of the voltage levels, admittances as well as the maximum thermal capacities of the transmission lines. The reactance of a line depends mainly on its physical properties. It increases proportionally to the geometric length of the line. Therefore, in the scope of this work, we assume equal physical properties for all lines and use the length to determine the reactance. A typical value for the reactance of a transmission line per unit length is 0.2 (Ω/km). Regarding the thermal capacities of the transmission lines, we assume a transmission capacity of 800 (MW) for 400 (kV) lines and 530 (MW) for 200 (kV) lines. In AC-OPF analysis the reactive power has strong influence on voltage drop thresholds. Thus, during AC OPF analysis, the maximum and minimum voltage levels for buses are considered and a value between 1.12 and 0.96 (p.u.) is assigned, respectively.

The gas network, comprises of 345 pipe segments with a total pipe length of roughly 4000 (km), 10 compressor stations and 352 nodes (54 exit stations to the local distribution system (CGS), 15 stations to direct served customers (14 GFPPs and one large industrial customer (IND)), two cross border export stations (CBE_1 & CBE_2), one cross border import station (CBI), one LNG terminal (LNG), one production field (PRO) and one underground gas storage facility (UGS). The CBI, PRO, LNG terminal and UGS facility are pressure controlled, while each compressor station is pressure ratio controlled with a pressure ratio set point ranging between 1.02 and 1.2. The input data for the compressor stations are listed in Table 3. The data used for the facilities supplying gas to the gas system are given in Table 4, while the data for the GFPPs are listed in Table 5. The minimum delivery pressure for the 14 GFPPs is set to 30 (bar-g) while the time needed to reach complete shut-down of a GFPP is set to 45 (min).

Table 3. Compressor station control (PRSET—Pressure Ratio Set Point) and constraints (PRMAX—Maximum Pressure Ratio, PWMAX—Maximum Available Driver Power, POMAX—Maximum Discharge Pressure, PIMIN—Minimum Suction Pressure).

Compressor Station	PRSET (-)	PRMAX (-)	PWMAX (MW)	POMAX (barg-g)	PIMIN (barg-g)
CS_1	1.05	1.6	10	54	34
CS_2	1.02	1.45	44	54	25
CS_3	1.01	1.6	60	54	25
CS_4	1.2	1.45	25	54	25
CS_5	1.2	1.45	80	54	25
CS_6	1.2	1.3	35	54	25
CS_7	1.2	1.45	50	54	25
CS_8	1.2	1.7	20	54	25
CS_9	1.2	1.7	20	54	25
CS_10	1.05	2	10	65	25

Table 4. Input data for facilities supplying the gas system with gas.

Gas Supply	k_0 (MW)	$k_1 \left(\frac{MW}{sm^3/s} \right)$	PSET (Barg)
CBI	-	-	50
PRO	-	-	52.6
UGS	3.5	0.01	56
LNG	5	0.03	50

Table 5. Input data for GFPPs connected to the gas and electric power system. Numbering of GFPPs corresponds to the numbering of the solid interconnection lines in Figure 4.

Name	c_0 (€)	$c_1 \left(\frac{€}{MW} \right)$	$c_2 \left(\frac{€}{MW^2} \right)$	η_T (%)	P_G^{max} (MW)	P_G^{min} (MW)	Q_G^{max} (MVAr)	Q_G^{min} (MVAr)	p^{min} (Barg)
GFPP_1	0	220.86	0	60	475	0	332.5	−285	30
GFPP_2	0	220.86	0	41	130	0	91	−78	30
GFPP_3	0	220.86	0	57	101	0	70.7	−61	30
GFPP_7	0	220.86	0	45	180	0	126	−108	30
GFPP_8	0	220.86	0	44.5	105	0	73.5	−63	30
GFPP_9	0	220.86	0	51	420	0	294	−252	30
GFPP_10	0	220.86	0	30	1127	0	788.9	−676	30
GFPP_11	0	220.86	0	40	360	0	252	−216	30
GFPP_12	0	220.86	0	48	420	0	294	−252	30
GFPP_14	0	220.86	0	30	766.7	0	536.7	−460	30
GFPP_15	0	220.86	0	45	147.8	0	103.5	−89	30
GFPP_16	0	220.86	0	61	435	0	304.5	−261	30
GFPP_17	0	220.86	0	67	390	0	273	−234	30
GFPP_18	0	220.86	0	55	410	0	287	−246	30

The transient scenarios for the integrated gas and power network are simulated by assigning the relative load profile depicted in Figure 5 to the relevant exit stations (left plot represents the gas load profile and right plot the power load profile). It should be noted that, the relative load profile for the gas system is only assigned to CGSs, which are the connection points between the gas transmission and local distribution system. For all other exit stations (CBE_1, CBE_2, IND) a constant load profile corresponding to the steady state load is assumed. The absolute values of the load profile for CGS nodes are obtained by multiplying the steady state load with the relative values in Figure 5 (left plot). The load profiles of the 14 GFPPs in the gas model are provided by the power model based on allocating the results obtained from the AC-OPF analysis to the corresponding nodes in the gas model. For the power network, the resulting loads for a time window of 24 h are obtained by multiplying the initial loads by the relative profile depicted in Figure 5 (right plot).

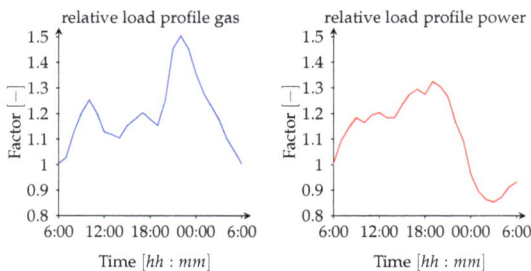

Figure 5. Load profiles gas (**left** side) and power (**right** side) networks.

All 14 GFPPs in the power grid are physically interconnected to the gas network. Furthermore, we assume additional interconnections between the gas and power network at two compressor stations, at the LNG terminal and at the UGS facility, which are supplied with power from the electric grid. The integrated gas and power network with 18 physically interconnected facilities is illustrated in Figure 4. Additional input parameters for the gas simulator are given in Table 6.

Table 6. Input data for the gas simulator.

Parameter	Symbol	Value	Unit
time step	Δt	900	(s)
total simulation time	t_{max}	24	(h)
gas temperature	T	288.15	(K)
dynamic viscosity	η	1.1×10^{-5}	(kg/m·s)
pipe roughness	k	0.012	(mm)
reference pressure	p_n	1.01325	(bar)
reference temperature	T_n	273.15	(K)
relative density	d	0.6	(-)
gross calorific value	GCV	41.215	(MJ/sm^3)

Applying the simulation tool SAInt on the presented sample network, some preliminary observations on cascading outage contingency analysis can be made. Initially, a base case scenario (scenario 0) with no supply disruption in any of the two interlinked networks is introduced. In the base case scenario, we capture the behavior of the networks at normal operation. Then, we compare the base case scenario with two scenarios, where we introduce a number of disruption events and simulate the reaction of the system to these events. The simulated grid is generated with a time resolution of 900 (s) and all scenarios are simulated for one gas day from 06:00 to 06:00 (For the case study, we chose a simulation time of one operating day (24 h) with a time resolution of 15 min for the gas model and a time resolution of one hour for the power model, in order to keep the size of input data and information at a moderate level for the results discussion. However, the framework is designed to allow an extension and adaptation of the time window and resolution depending on if a short or long term study of a contingency scenario is of interest.) It should be noted that although it is possible to change the status of the failed components (repairing and restoration can be modeled) within the simulation, the scenarios that are presented in this study do not take into account the repairing activity in order to analyze system capabilities in worst-cases.

While the first scenario involves a disruption of several supply points in the gas network, the second scenario includes supply disruptions triggered by the power network. In scenario 1, we assume a reduced regasification rate for the LNG terminal from maximum via a ramp-down between 06:00 and 07:00 (see Figure 6), which corresponds to an expected 7-day delay in cargo. In addition, we assume a supply disruption at the production field causing a ramp down of the supply between 08:00 and 9:00 (see Figure 6). Furthermore, a 30% supply reduction at CBI station at time 14:00 is implemented via a ramp-down between 14:00 and 15:00 (see Figure 6). Scenario 2 is related to power network contingencies and initial contingency set consist of the loss of major lignite power plant with 1157 (MW) operational capacity at 07:00 and 70% lack of power generation from wind turbines at 06:00 (see Figure 6).

In the following, we discuss the simulation results for the three scenarios (The simulation results and conclusions are based on the input data chosen for the sample network. While some data were provided by the TSOs, others were not available (e.g., pipe roughness, gas temperature, line properties etc.) and were therefore estimated using typical values. Thus, these input data are connected with uncertainties).

The sequence of initial events (shown in black) and their consequences (shown in orange and red) are summarized in Figures 6 and 7 for scenario 1 and scenario 2, respectively. It can be seen from the figures, that when a minimum pressure violation for a GFPP is detected in the gas model, the failure of the corresponding power plant is applied after 45 (min) due to the required shut-down time. Figures 8–10 show the difference in gas supply to the system through the CBI station, the production field and the LNG terminal. There is a big difference in inflows to the system through these supply points in scenario 0 and scenario 1, where the difference is more than 20 (Msm^3/d) (Million standard cubic meter per day, where the reference pressure is 1.0135 (bar) and the reference temperature is 0 (°C)). The impact of this observation can be seen in Figures 8–12. Figure 11 shows that the disruptions introduced in scenario 1 have the highest impact on the gas network, since the flow balance, which is the sum of inflow minus sum of outflow, is always negative; the system is not able to supply enough gas to balance the demand. In fact, the flow balance is quite negative throughout the time, peaking down to equivalent daily flows of −32 (Msm^3/d). As a result, the quantity of gas stored in the pipeline (i.e., the line pack) reduces significantly as time passes. The flow balance can be viewed as the time derivative of the line pack, thus, if the flow balance is negative the line pack decreases and if positive the line pack increases. A zero flow balance corresponds to no change in line pack. Latter is the assumption made in steady state gas models, which cannot capture the changes in line pack, and therefore, the real behavior of the gas system appropriately. Moreover, Figure 11 shows a decrease in line pack from ca. 85 to 67 (Msm^3/d) for scenario 1 (approx. 18 (Msm^3/d) lost along the day in the pipelines). In contrast, in scenario 0 only approx. 1.5 (Msm^3/d) of line pack is extracted.

This produces a steady decrease of pressure in the CBI station, the production field and the LNG facility causing the pressure to reduce to approx. 39, 42 and 31 (bar-g), respectively (see Figures 8–10).

An important observation is the pressure drop to approximately 31 (bar-g) at the LNG terminal, which is the main gas supplier for some of the GFPPs in the hydraulic region. This value is slightly above the 30 (bar-g) minimum delivery pressure threshold required by the GFPPs. When gas supplies are scarce, the only way to keep maintain sufficient pressure and to allow the network to continue operating is to reduce consumption, either through curtailment or fuel switching, if there is the chance to do this with some power plants. In scenario 1, gas curtailment at GFPPs is implemented, presuming that replacement fuel is not available in any of the investigated GFPPs.

Figure 12 shows the behavior of the UGS facility, the only supply node able to increase gas supply to satisfy the increased demand in scenario 1. The UGS facility is able to maintain its pressure set point till the end of the simulation (see Figure 12). The disconnection of four GFPPs from the gas network at 14:15, 15:45 and 16:30, respectively, allows the gas system to continue running (see Figures 6 and 13).

The pressure and load profiles for failed GFPPs are given in Figures 13 and 14. This curtailment was sufficient to cope with the pressure drop in the network. Therefore, there was no need of gas curtailment at CGSs, where protected customers (e.g., households, public services) are supplied with gas.

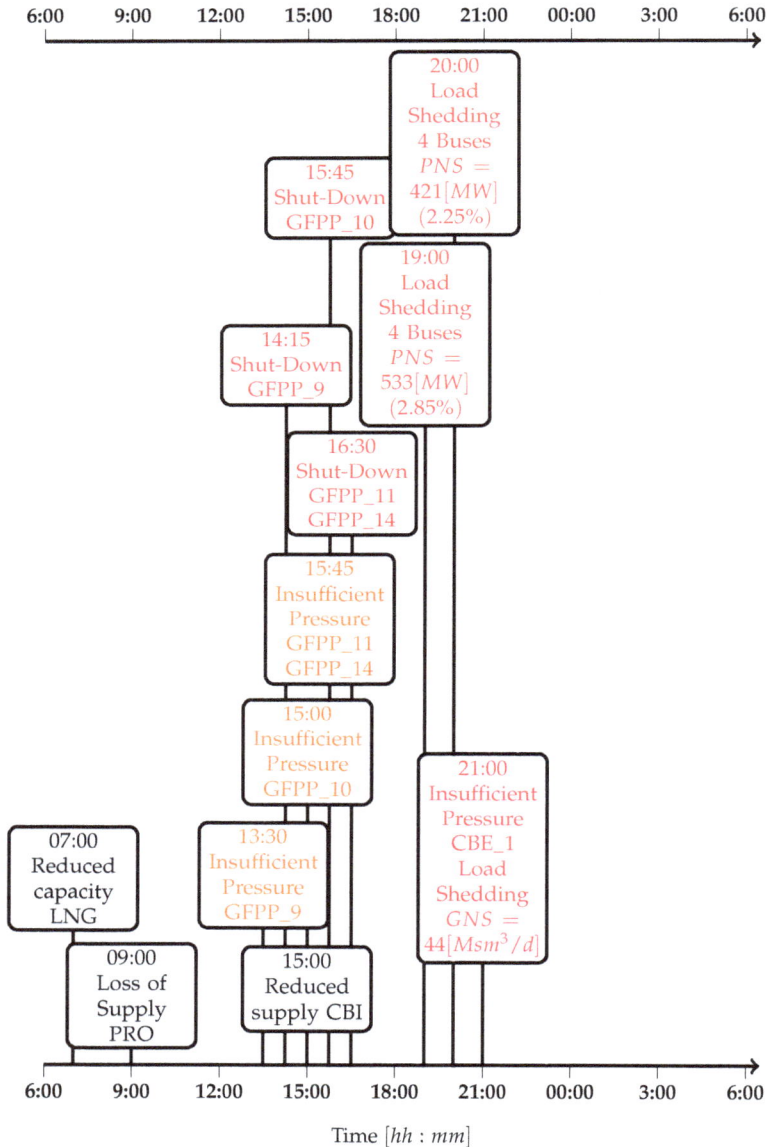

Figure 6. Timing of initial (**black**) and cascading (**orange**, **red**) events for Scenario 1. Abbreviation PNS stands for power not supplied, while GNS stands for gas not supplied, value in brackets refers to the fraction of not supplied power/gas with respect to total power/gas loads.

Figures 15 and 16 depicts the voltage profiles for a selected number of buses, where minimum voltage violation is detected for scenario 1 and 2, respectively. In order to keep the bus voltage above the minimum voltage level, load shedding is implemented at the affected buses. The left plots in Figures 15 and 16 show the voltage profiles of the affected buses for the computation where voltage violations were detected and no countermeasures were employed to avoid this violation, while the right plots show the voltage profiles after implementing load shedding at the affected buses. As can be seen in the right plots of Figures 15 and 16, the bus voltages recover to a value above the minimum voltage threshold after load shedding is implemented. However, due to load shedding some customers connected to the affected buses are not supplied with enough electricity (see Figures 6 and 7).

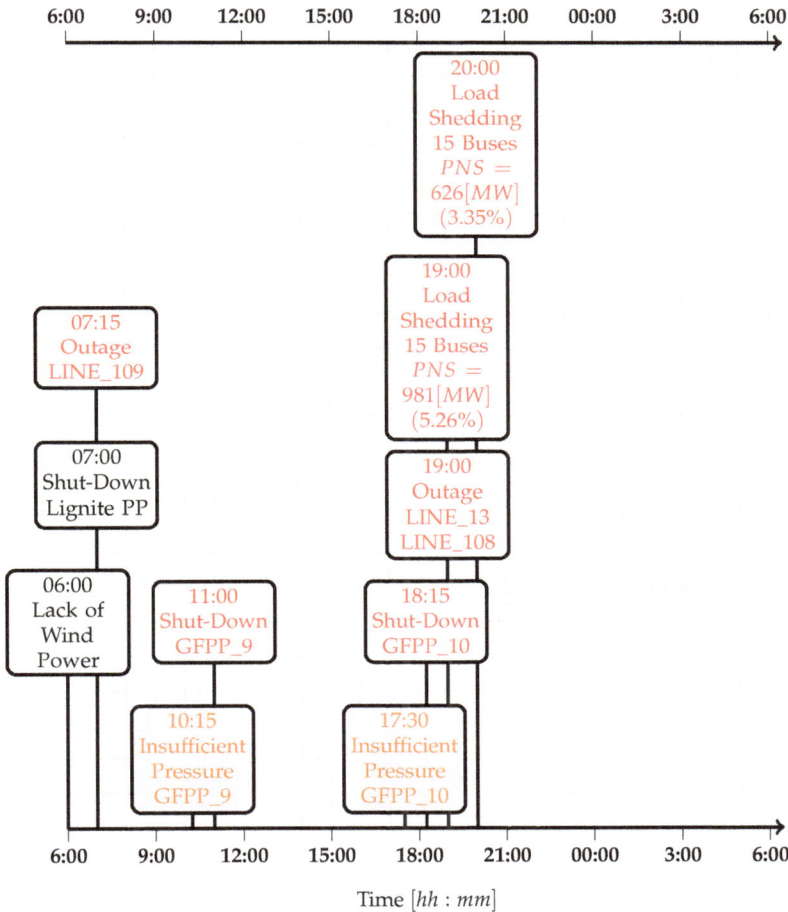

Figure 7. Timing of initial (**black**) and cascading (**orange, red**) events for Scenario 2. Abbreviation PNS stands for power not supplied, value in brackets refers to the fraction of not supplied power with respect to total loads.

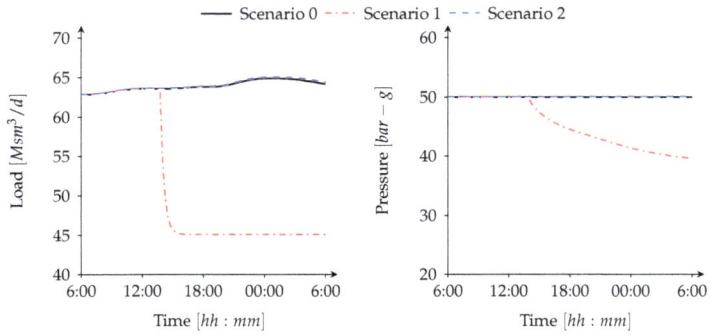

Figure 8. Time evolution of gas supply and pressure at the cross border import (CBI) node for the computed scenarios

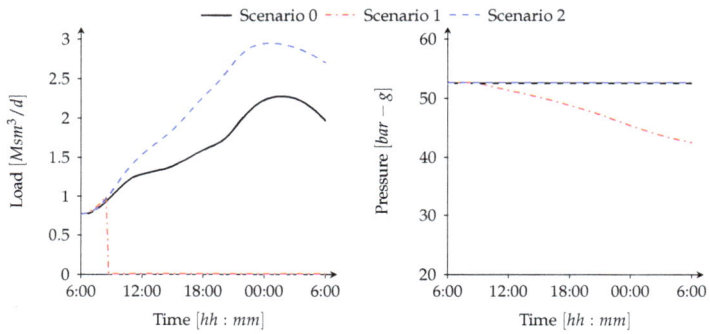

Figure 9. Time evolution of gas supply and pressure at the production field for the computed scenarios.

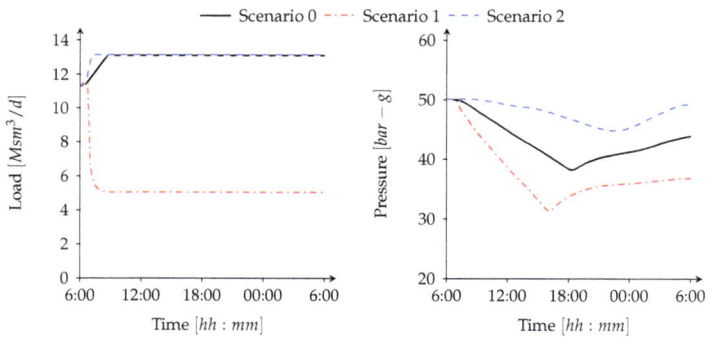

Figure 10. Time evolution of regasification rate and pressure at the liquefied natural gas (LNG) terminal for the computed scenarios.

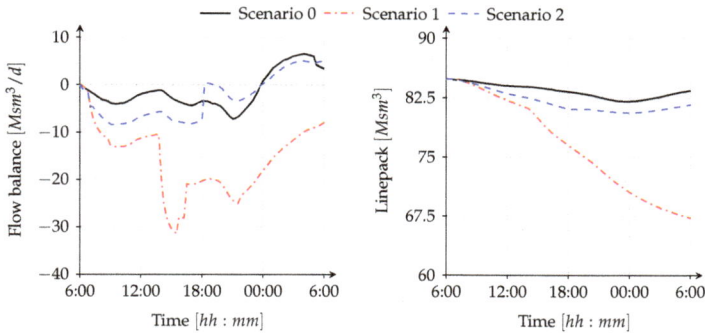

Figure 11. Time evolution of flow balance (sum of inflow minus sum of outflow) and line pack for the computed scenarios.

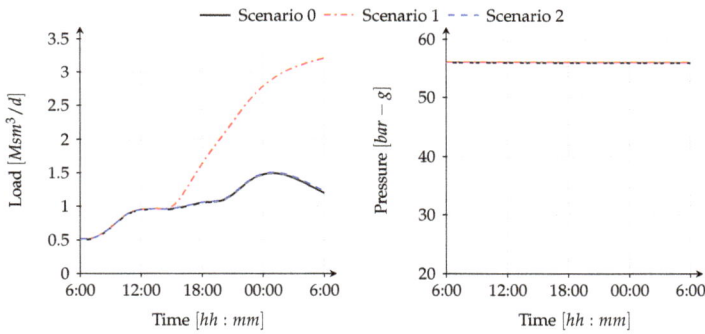

Figure 12. Time evolution of withdrawal rate and pressure at underground gas storage (UGS) facility for the computed scenarios.

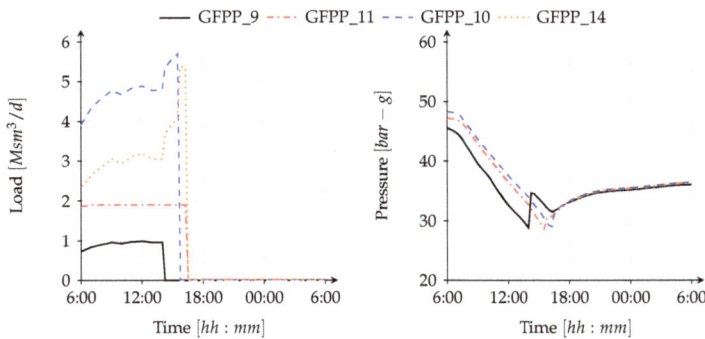

Figure 13. Time evolution of load and pressure of failed GFPPs in scenario 1.

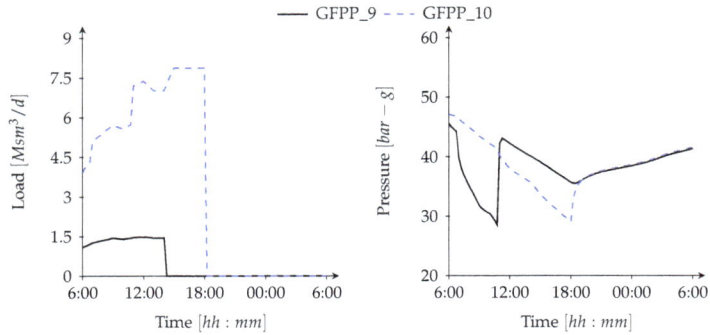

Figure 14. Time evolution of load and pressure of failed GFPPs in scenario 2.

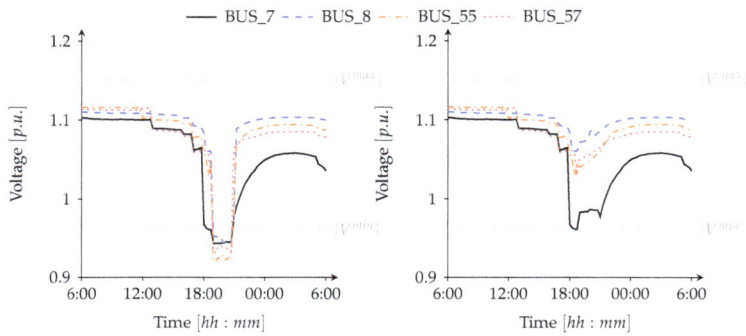

Figure 15. Time evolution of bus voltage before load shedding (**left**) and after (**right**) for scenario 1. All four buses where load shedding was applied are shown in this figure.

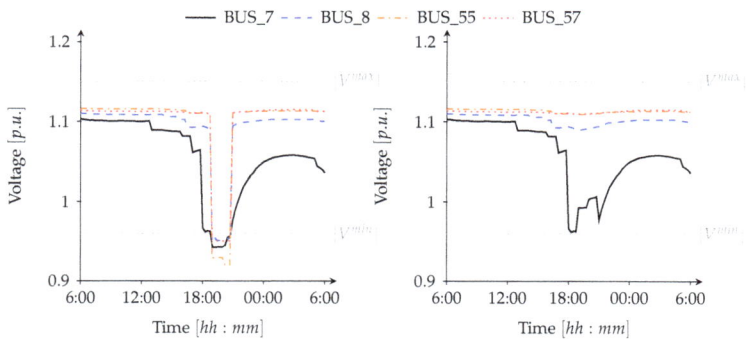

Figure 16. Time evolution of bus voltages before load shedding (**left**) and after (**right**) for scenario 2. Load shedding was applied at 15 buses. Among these buses are the 4 buses from scenario 1, which are shown in this figure.

Regarding the CBE_1 station, due to the pressure drop at the station (see Figure 17), the flow is restricted around 21:00 because the threshold pressure of 30 (bar-g) is reached. This is the only way to keep minimum delivery pressure at that exit point; otherwise problems would arise downstream due to too low pressure. Figure 17, shows the drop in flow (around 8 (Msm^3/d)) at CBE_1 station due to the pressure restriction.

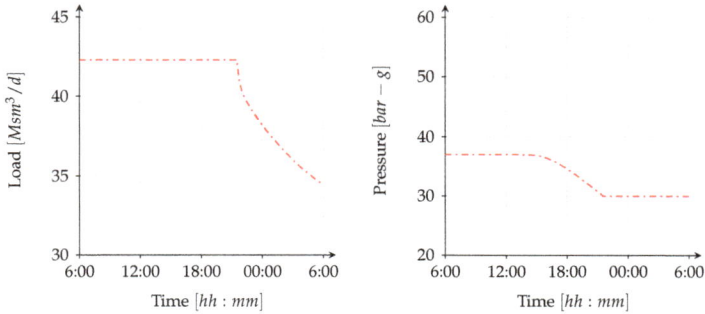

Figure 17. Load and pressure profile of CBE_1 for scenario 1.

Moreover, the difference between scenario 0 and scenario 2 shows the gas system reaction to the electric side disruption. In Figure 11, it can be seen that the flow balance of the gas network in scenario 2 is more negative (the gas system loses more gas) than in scenario 0 until 18:00. This is caused by the increase in gas demand of GFPPs due to the disruption of the lignite power plant and the loss of power generation from wind turbines. The increase in gas demand of GFPPs leads to a pressure drop in two GFPPs, followed by the disconnection of the power plants from the network (see Figures 7 and 14). The pressure and load profiles for failed GFPPs are given in Figure 14. The disconnection of the generators affects the loading of the gas system in a positive way. Moreover, the line pack starts to recover after 18:00 (see Figure 11).

The scenario results indicate clearly that the disruptions taking place in the gas network that affect GFPPs also affected the operability of the power network. After failure of each GFPP, the power model calculates the new generating profiles for all power plants and sends these profiles to the gas model. In scenario 1, the closure of 4 GFPPs due to low pressure levels in the gas system caused voltage violations in the electricity network at peak demand hour (19:00–20:00) because of the high amount of power transmission from relatively distant generators in order to compensate the missing generating capacity. This violation in voltage levels caused 954 (MW) of load shedding during 2 h (see Figures 6 and 15). In scenario 2, the cascading effects are more severe including three line overloads and load shedding of 1607 (MW) at the peak demand hours (19:00–20:00, see Figures 7 and 16). The initial failure of large capacity lignite power plant together with lack of power generation from wind power caused an increase in power generation from GFPPs. This increase results in pressure drops at two GFPPs followed by the closure of both facilities. The system has to implement these cascading effects in order to avoid a complete blackout in the overall network.

Furthermore, the results show that the impact of disruptions introduced in both scenarios is much higher for the power system than for the gas system Section 3.

6. Conclusions

In this paper, we developed an integrated simulation framework for cascading outage contingency analysis in combined gas and power system networks and demonstrated the capabilities of the implemented framework by applying it to a realistic, combined electricity and gas transmissions network of an European region.

The simulation framework is composed of a transient hydraulic model for the gas system and a steady state AC-OPF model for the power system. Both models, are derived from the physical laws governing the flow of gas and electrical power, respectively. Moreover, the most important facilities and their technical constraints are considered. The gas and power system models a coupled through coupling equations describing the fuel gas offtake of GFPPs for power generation and the power supply to LNG terminals, UGS facilities and electric driven compressor stations.

The model application was divided into three scenarios, namely, scenario 0 with no disruption, scenario 1 with gas side disruptions and scenario 2 with power side disruptions. The results of these scenarios show how disruption events triggered in one system propagate to the other system. In scenario 1, for instance, three major gas supply stations are disrupted and as a result a number of GFPPs are shut-down due to insufficient fuel gas pressure. This contingency propagates further to other buses in the power system, where load shedding is implemented in order to maintain the voltage levels above the minimum voltage threshold. Similar observations are made in scenario 2, where a drastic reduction in renewable energy generation together with a shutdown of a large power plant triggered a large increase in gas demand of GFPPs, leading to a rapid pressure drop in the gas network and the subsequent shut-down of GFPPs. Eventually, this circumstance increased the stress on the power system leading to minimum bus voltage violations in a couple of buses, which is remedied by applying load shedding at the affected buses.

Based on these key findings, it can be concluded that there is a need for close collaboration and coordination between gas and power TSOs. Data concerning pressures, flows, voltages etc., efficiently handled and communicated may introduce resilience on the integrated network. This has to be done via well-structured protocols that inform the other TSO about the grace periods and support that each network may grant the other. The use of models like the one proposed in this study may be of much help for getting part of this information to share with the other operator.

We believe it is fair to state that the integrated model allows for detailed fingerprinting and exploration of the effects of disruption in gas and/or power, to a level of detail that is not possible by qualitative, expert analysis. Once the data characterizing a gas and electricity grid have been loaded, experts can perform in-silico experiments at will to investigate the system, determine weak elements, and propose mitigation strategies. In both two scenarios, GFPP_9 and GFPP_10 fail, which merits an investigation into their position in the system. If in more scenarios it is these two plants that fail first, it could be decided to equip these with alternative backup fuel options. In the future, we intend to further develop the simulation framework to implement more simulation options and functionalities into the simulation tool SAInt in order to investigate the effectiveness of different demand and supply side measures to mitigate the consequences of supply disruptions in coupled gas and electric power systems.

Acknowledgments: This work has been developed within the framework of the European Program for Critical Infrastructure Protection (EPCIP) of the European Commission. We express our gratitude to our colleague Nicola Zaccarelli from the Joint Research Centre—Institute for Energy and Transport, in Petten, Netherlands, for providing the GIS-Data for the presented gas and electric model. We would also like to thank Tom van der Hoeven for the productive discussions and suggestions, which have improved the simulation tool SAInt.

Author Contributions: Kwabena Addo Pambour developed the simulation software SAInt, designed and implemented the simulation framework into SAInt and wrote the paper; Burcin Cakir-Erdener designed the simulation framework and the case studies, performed the computations, analyzed the results, and wrote the paper; Ricardo Bolado-Lavin reviewed the paper and proposed modifications to improve the design of the simulation framework and the case studies; Gerard P. J. Dijkema reviewed the paper and suggested changes to improve the quality of the paper.

Conflicts of Interest: The authors declare no conflict of interest.

Abbreviations

The following abbreviations are used in this manuscript:

Abbreviations

AC	Alternating current
API	Application Programming Interface
EU	European Union

ED	Economic dispatch
CBE	Cross Border Export
CBI	Cross Border Import
CBP	Cross Border Point
CEI	Critical Energy Infrastructures
CGS	City Gate Station
DC	Direct current
DTA	Dynamic Time Step Adaptation
GB	Great Britain
GFPP	Gas Fired Power Plant
GNS	Gas not supplied
GUI	Graphical User Interface
IND	Large Industrial Customer
KCL	Kirchoff's Current Law
LNG	Liquefied Natural Gas
NGTS	National Gas Transport System
P2G	Power to Gas
PF	Power Flow
PDE	Partial Differential
PNS	Power not supplied
PRO	Production Fields
OPF	Optimal Power Flow
SAInt	Scenario Analysis Interface
SCUC	Security Constraint Unit commitment
SNG	Synthetic Natural Gas
TSO	Transmission System Operator
UC	Unit commitment
UGS	Underground Gas Storage

Mathematical Symbols

A	cross-sectional area
a	transformer tap ratio
$a_{i,j}$	elements of the node-branch incidence matrix
b	line charging susceptance
c_0, c_1, c_2	coefficients of cost function
c	speed of sound
CV	control volume
D	inner pipe diameter
e	Euler's number
E	set of branches
f	electric driver factor
g	gravitational acceleration
G	directed graph
GCV	gross calorific value
I_f	electric curent injection at from bus
I_t	electric curent injection at to bus
j	imaginary number
k_0, k_1	coefficients of coupling equation
L	nodal load
L_{GFPP}	fuel gas offtake for power generation at GFPPs
l	pipe length
LP	line pack
M	number of pipe section
n	simulation time point
N_n	number of gas nodes

N_b	number of buses, number of branches
N_{CS}	number of compressor stations
N_g	number of power generation units
N_{GFPP}	number of GFPPs
N_{iq}	number of inequality constraints
N_l	number of transmission lines and transformers
P_D	active power demand
P_D^{CS}	power demand of compressor stations
P_D^{rw}	power demand of LNG terminals and UGS facilities
P_G	active power generation
$\mathbf{P_G}$	vector of active power generation
p	gas pressure (vector)
p_1	inlet pressure
p_2	outlet pressure
p_m	mean pipe pressure
Q	gas flow rate, reactive power
$\mathbf{Q_G}$	vector of reactive power generation
R	gas constant, line resistance
S_k^f	apparent power injection at from bus of branch k
S_k^{max}	maximum transmission capacity of branch k
S_k^t	apparent power injection at to bus of branch k
$\mathbf{S}$	vector of apparent power flow
t	time, complex transformer tap
t_n	time point
Δt	time step
T	temperature
T_n	reference temperature
v	gas velocity
V	complex bus voltage, set of nodes
$\mathbf{V}$	vector of complex bus voltage
$\mathbf{V_m}$	vector of complex bus voltage magnitudes
$|V|$	bus voltage magnitude
V_i	nodal volume
X	line reactance
x	pipeline coordinate
$\mathbf{X}$	vector of decision variables
Δx	pipe segment length
Y	line admittance
$\mathbf{Y_{br}}$	branch admittance matrix
$\mathbf{Y_{bus}}$	bus admittance matrix
Z	compressibility factor, impedance

Greek Symbols

α	inclination
α, β, γ	coefficients of heat rate curve
δ	voltage angle
Δ	vector of bus voltage angles
ϵ	residual tolerance
η_{ad}	compressor adiabatic efficiency
η_m	driver efficiency
η_T	thermal efficiency
κ	isentropic exponent
λ	friction factor

ϕ	transformer phase shift angle
ρ	gas density
ρ_n	gas density at reference conditions

Physical Units

(bar-g)	bar gauge (absolute pressure minus atmospheric pressure)
(p.u.)	per unit
(Msm^3)	millions of standard cubic meters (line pack, inventory)
(Msm^3/d)	millions of standard cubic meters per day (gas flow rate)
(sm^3)	standard cubic meters (line pack, inventory)

References

1. North American Electric Reliability Corporation (NERC). *Accommodating an Increased Dependence on Natural Gas for Electric Power, Phase II: A Vulnerability and Scenario Assessment for the North American Bulk Power*; Technical Report; North American Electric Reliability Corporation (NERC): Atlanta, GA, USA, 2013.
2. Pearson, I.; Zeniewski, P.; Gracceva, F.; Zastera, P.; McGlade, C.; Sorrell, S. *Unconventional Gas: Potential Energy Market Impacts in the European Union*; Technical Report; JRC Scientific and Policy Reports EUR 25305 EN; Joint Research Centre of the European Commission: Petten, The Netherlands, 2012.
3. Judson, N. *Interdependence of the Electricity Generation System and the Natural Gas System and Implications for Energy Security*; Technical Report; Massachusetts Institute of Technology Lincoln Laboratory: Lexington, MA, USA, 2013.
4. Clegg, S.; Mancarella, P. Integrated modeling and assessment of the operational impact of power-to-gas (P2G) on electrical and gas transmission networks. *IEEE Trans. Sustain. Energy* **2015**, *6*, 1234–1244.
5. Acha, S. Impacts of Embedded Technologies on Optimal Operation of Energy Service Networks. Ph.D. Thesis, University of London, London, UK, 2010.
6. Rubio Barros, R.; Ojeda-Esteybar, D.; Ano, A.; Vargas, A. Integrated natural gas and electricity market: A survey of the state of the art in operation planning and market issues. In Proceedings of the Transmission and Distribution Conference and Exposition: Latin America 2008 IEEE/PES, Bogota, Colombia, 13–15 August 2008.
7. Rubio Barros, R.; Ojeda-Esteybar, D.; Ano, A.; Vargas, A. Combined operational planning of natural gas and electric power systems: State of the art. In *Natural Gas*; Potocnik, P., Ed.; INTECH Open Access Publisher: Rijeka, Croatia, 2010.
8. Cakir Erdener, B.; Pambour, K.A.; Lavin, R.B.; Dengiz, B. An integrated simulation model for analysing electricity and gas systems. *Int. J. Electr. Power Energy Syst.* **2014**, *61*, 410–420.
9. Abrell, J.; Weigt, H. *Combining Energy Networks*; Electricity Markets Working Papers WP-EM-38 2010; TU Dresden: Dresden, Germany, 2010.
10. Özdemir, O.; Veum, K.; Joode, J.; Migliavacca, G.; Grassi, A.; Zani, A. The impact of large-scale renewable integration on Europe's energy corridors. In Proceedings of the IEEE Trondheim PowerTech, Trondheim, Norway, 19–23 June 2011; pp. 1–8.
11. Möst, D.; Perlwitz, H. Prospect of gas supply until 2020 in Europe and its relevance of the power sector in the context of emissions trading. *Energy* **2008**, *34*, 1510–1522.
12. Lienert, M.; Lochner, S. The importance of market interdependencies in modelling energy systems—The case of European electricity generation market. *Int. J. Electr. Power Energy Syst.* **2012**, *34*, 99–113.
13. Chen, H.; Baldick, R. Optimizing short-term natural gas supply portfolio for electric utility companies. *IEEE Trans. Power Syst.* **2007**, *22*, 232–239.
14. Asif, U.; Jirutitijaroen, P. An optimization model for risk management in natural gas supply and energy portfolio of a generation company. In Proceedings of TENCON 2009—2009 IEEE Region 10 Conference, Singapore, 23–26 January 2009.
15. Kittithreerapronchai, O.; Jirutitijaroen, P.; Kim, S.; Prina, J. Optimizing natural gas supply and energy portfolios of a generation company. In Proceedings of PMAPS 2010—IEEE 11th International Conference on Probabilistic Methods Applied to Power Systems, Singapore, 14–17 June 2010; pp. 231–237.

16. Duenas, P.; Barquin, J.; Reneses, J. Strategic management of multi-year natural gas contracts in electricity markets. *IEEE Trans. Power Syst.* **2012**, *27*, 771–779.

17. Spiecker, S. Analyzing market power in a multistage and multiarea electricity and natural gas system. In Proceedings of EEM—8th IEEE International Conference on the European Energy Market, Zagreb, Croatia, 25–27 May 2011; pp. 313–320.

18. Weigt, H.; Abrell, J. Investments in a combined energy network model: Substitution between natural gas and electricity. In Proceedings of the Annual Conference 2014 (Hamburg): Evidence-Based Economic Policy, Hamburg, Germany, 7–10 September 2014.

19. Gil, M.; Dueñas, P.; Reneses, J. Electricity and natural gas interdependency: Comparison of two methodologies for coupling large market models within the European regulatory framework. *IEEE Trans. Power Syst.* **2016**, *31*, 361–369.

20. Shahidehpour, M.; Fu, Y.; Wiedman, T. Impact of natural gas infrastructure on electricity power system security. *IEEE Proc.* **2005**, *93*, 1042–1056.

21. Li, T.; Eremia, M.; Shahidehpour, M. Interdependency of natural gas network and power system security. *IEEE Trans. Power Syst.* **2008**, *23*, 1817–1824.

22. Liu, C.; Shahidehpour, M.; Fu, Y.; Li, Z. Security-constrained unit commitment with natural gas transmission constraints. *IEEE Trans. Power Syst.* **2009**, *24*, 1523–1536.

23. Li, G.; Zhang, R.; Chen, H.; Jiang, T.; Jia, H.; Mu, Y.; Jin, X. Security-constrained economic dispatch for integrated natural gas and electricity systems. *Energy Procedia* **2016**, *88*, 330–335.

24. An, S.; Li, Q.; Gedra, T. Natural gas and electricity optimal power flow. *IEEE Transm. Distrib. Conf. Expo.* **2003**, *1*, 138–143.

25. Unsihuay, C.; Marangon Lima, J.; Zambroni de Souza, A. Modelling the integrated natural gas and electricity optimal power flow. In Proceedings of IEEE Power Engineering Society General Meeting, Tampa, FL, USA, 24–28 June 2007.

26. Martínez-Mares, A.; Fuerte-Esquivel, C. Integrated energy flow analysis in natural gas and electricity coupled systems. In Proceedings of IEEE NAPS—North American Power Symposium, Boston, MA, USA, 4–6 August 2011; pp. 1–7.

27. Liu, C.; Shahidehpour, M.; Wang, J. Coordinated scheduling of electricity and natural gas infrastructures with a transient model for natural gas flow. *Chaos Interdiscip. J. Nonlinear Sci.* **2011**, *21*, doi:10.1063/1.3600761.

28. Chaudry, M.; Jenkins, N.; Strbac, G. Multi-time periode combined gas and electricity network optimisation. *Electr. Power Syst. Res.* **2008**, *78*, 1265–1279.

29. Clegg, S.; Mancarella, P. Integrated electrical and gas network modelling for assessment of different power-and-heat options. In Proceedings of the Power Systems Computation Conference (PSCC), Wroclaw, Poland, 18–22 August 2014; pp. 1–7.

30. Qadrdan, M.; Chaudry, M.; Wu, J.; Jenkins, N.; Ekanayake, J. Impact of a large penetration of wind generation on the GB gas network. *Energy Policy* **2010**, *38*, 5684–5695.

31. Qadrdan, M.; Jenkins, N.; Ekanayake, J. Operating strategies for a GB integrated gas and electricity network considering the uncertainty in wind power forecasts. *IEEE Trans. Sustain. Energy* **2014**, *5*, 128–138.

32. Chertkov, M.; Backhaus, S.; Lebedev, V. Cascading of fluctuations in interdependent energy infrastructures: Gas-grid coupling. *Appl. Energy* **2015**, *160*, 541–551.

33. Alabdulwahab, A.; Abusorrah, A.; Zhang, X.; Shahidehpour, M. Coordination of Interdependent Natural Gas and Electricity Infrastructures for firming the variability of wind energy in stochastic day-ahead scheduling. *IEEE Trans. Sustain. Energy* **2015**, *6*, 606–615.

34. Cui, H.; Li, F.; Hu, Q.; Bai, L.; Fang, X. Day-ahead coordinated operation of utility-scale electricity and natural gas networks considering demand response based virtual power plants. *Appl. Energy* **2016**, *176*, 183–195.

35. Bai, L.; Li, F.; Cui, H.; Jiang, T.; Sun, H.; Zhu, J. Interval optimization based operating strategy for gas-electricity integrated energy systems considering demand response and wind uncertainty. *Appl. Energy* **2016**, *167*, 270–279.

36. Zhang, X.; Shahidehpour, M.; Alabdulwahab, A.; Abusorrah, A. Hourly electricity demand response in the stochastic day-ahead scheduling of coordinated electricity and Natural Gas Networks. *IEEE Trans. Power Syst.* **2016**, *31*, 592–601.

37. Qadrdan, M.; Cheng, M.; Wu, J.; Jenkins, N. Benefits of demand-side response in combined gas and electricity networks. *Appl. Energy* **2016**, doi:10.1016/j.apenergy.2016.10.047.

38. Saldarriaga, C.A.; Hincapié, R.A.; Salazar, H. A holistic approach for planning natural gas and electricity distribution networks. *IEEE Trans. Power Syst.* **2013**, *28*, 4052–4063.

39. Chaudry, M.; Jenkins, N.; Qadrdan, M.; Wu, J. Combined gas and electricity network expansion planning. *Appl. Energy* **2014**, *113*, 1171–1187.

40. Qadrdan, M.; Abeysekera, M.; Chaudry, M.; Wu, J.; Jenkins, N. Role of power-to-gas in an integrated gas and electricity system in Great Britain. *Int. J. Hydrogen Energy* **2015**, *40*, 5763–5775.

41. Clegg, S.; Mancarella, P. Storing renewables in the gas network: Modelling of power-to-gas seasonal storage flexibility in low-carbon power systems. *IET Gener. Transm. Distrib.* **2016**, 10, 566–575.

42. Urbina, M.; Li, Z. A combined model for analysing the interdependency of electrical and gas systems. In Proceedings of NAPS—39th IEE North American Power Symposium, Las Cruces, NM, USA, 30 September–2 October 2007.

43. Urbina, M.; Li, Z. Modeling and analysing the impact of interdependency between natural gas and electricity infrastructures. In Proceedings of 2008 IEEE Power and Energy Society General Meeting—Conversion and Delivery of Electrical Energy in the 21st Century, Pittsburgh, PA, USA, 20–24 July 2008.

44. Carvalho, R.; Buzna, L.; Bono, F.; Gutierrez, E.; Just, W.; Arrowsmith, D. Robustness of trans-European gas networks. *Phys. Rev. E* **2009**, *80*, 016106.

45. Munoz, J.; Jimenez-Redondo, N.; Perez-Ruiz, J.; Barquin, J. Natural gas network modeling for power systems reliability studies. In Proceedings of the 2003 IEEE Bologna Power Tech Conference, Bologna, Italy, 23–26 June 2003; p. 6.

46. Arnold, M.; Andersson, G. Decomposed electricity and natural gas optimal power flow. In Proceedings of the 16th Power Systems Computation Conference (PSCC), Glasgow, Scotland, 14–18 July 2008.

47. Jaworsky, C.; Spataru, C.; Turitsyn, K. Vulnerability assessment of Interdependent gas and electricity networks. In Proceedings of the Hawaii International Conference on System Sciences (HICSS48), Kauai, HI, USA, 5–8 January 2015.

48. Osiadacz, A.J. Different transient flow models-limitations, advantages, and disadvantages. In Proceedings of the Pipeline Simulation Interest Group (PSIG) Annual Meeting, San Francisco, CA, USA, 24–25 October 1996.

49. Clegg, S.; Mancarella, P. Integrated electrical and gas network flexibility assessment in low-carbon multi-energy systems. *IEEE Trans. Sustain. Energy* **2016**, *7*, 718–731.

50. Zlotnik, A.; Chertkov, M.; Carter, R.; Hollis, A.; Daniels, A.; Backhaus, S. Using power grid schedules in dynamic optimization of gas pipelines. In Proceedings of the Pipeline Simulation Interest Group (PSIG) Conference 2016, Vancouver, BC, Canada, 10–13 May 2016.

51. Zimmerman, R.D.; Murillo-Sanchez, C.E.; Thomas, R.J. MATPOWER: Steady-state operations, planning and analysis tools for power systems research and education. *IEEE Trans. Power Syst.* **2011**, *26*, 12–19.

52. Pambour, K.A.; Cakir Erdener, B.; Bolado-Lavin, R.; Dijkema, G.P.J. An integrated simulation tool for analyzing the operation and interdependency of natural gas and electric power systems. In Proceedings of the Pipeline Simulation Interest Group (PSIG) Conference 2016, Vancouver, BC, Canada, 10–13 May 2016.

53. Pambour, K.A.; Bolado-Lavin, R.; Dijkema, G.P. An integrated transient model for simulating the operation of natural gas transport systems. *J. Nat. Gas Sci. Eng.* **2016**, *28*, 672–690.

54. Pambour, K.A.; Bolado-Lavin, R.; Dijkema, G.P. SAInt—A simulation tool for analysing the consequences of natural gas supply disruptions. In Proceedings of the Pipeline Technology Conference (PTC) 2016, Berlin, Germany, 23–25 May 2016.

55. Van der Hoeven, T. *Math in Gas and the art of linearization*; Energy Delta Institute: Groningen, The Netherlands, 2004.

56. Cerbe, G. *Grundlagen der Gastechnik*; Carl Hanser Verlag: München, Germany, 2008.

57. Osiadacz, A.J. *Simulation and Analysis of Gas Networks*; Gulf Publishing Company: Houston, TX, USA, 1987.

58. Carreras, B.A.; Lynch, V.E.; Dobson, I.; Newman, D.E. Critical points and transitions in an electric power transmission model for cascading failure blackouts. *Chaos Interdisc. J. Nonlin. Sci.* **2002**, *12*, 985–994.

59. Baldick, R.; Chowdhury, B.; Dobson, I.; Dong, Z.; Gou, B.; Hawkins, D.; Huang, H.; Joung, M.; Kirschen, D.; Li, F.; et al. Initial review of methods for cascading failure analysis in electric power transmission systems IEEE PES CAMS task force on understanding, prediction, mitigation and restoration of cascading failures. In Proceedings of the 2008 IEEE Power and Energy Society General Meeting—Conversion and Delivery of Electrical Energy in the 21st Century, Pittsburgh, PA, USA, 20–24 July 2008; pp. 1–8.

60. Papic, M.; Bell, K.; Chen, Y.; Dobson, I.; Fonte, L.; Haq, E.; Hines, P.; Kirschen, D.; Luo, X.; Miller, S.S.; et al. Survey of tools for risk assessment of cascading outages. In Proceedings of the 2011 IEEE Power and Energy Society General Meeting, Detroit, MI, USA , 24–28 July 2011; pp. 1–9.

61. Ren, H.; Fan, X.; Watts, D.; Lv, X. Early warning mechanism for power system large cascading failures. In Proceedings of the 2012 IEEE International Conference on Power System Technology (POWERCON), Auckland, New Zealand, 23–26 October 2012; pp. 1–6.

62. Vaiman, M.; Bell, K.; Chen, Y.; Chowdhury, B.; Dobson, I.; Hines, P.; Papic, M.; Miller, S.S.; Zhang, P. Risk assessment of cascading outages: Part I 2014; Overview of methodologies. In Proceedings of the 2011 IEEE Power and Energy Society General Meeting, Detroit, MI, USA , 24–28 July 2011; pp. 1–10.

63. Vaiman, M.; Bell, K.; Chen, Y.; Chowdhury, B.; Dobson.; Hines, P.; Papic, M.; Miller, S.S.; Zhang, P. Risk assessment of cascading outages: Methodologies and challenges. *IEEE Trans. Power Syst.* **2012**, *27*, 631–641.

64. Mozina, C.J. Power plant protection and control strategies for blackout avoidance. In Proceedings of the 2006 Power Systems Conference on Advanced Metering, Protection, Control, Communication, and Distributed Resources, Clemson, South Carolina, 14–17 March 2006.

65. North American Electric Reliability Council (NERC). *System Disturbance Report*; Technical Report; North American Electric Reliability Council: Atlanta, GA, USA, 1998.

66. Begovic, M.; Fulton, D.; Gonzalez, M.R.; Goossens, J.; Guro, E.A.; Haas, R.W.; Henville, C.F.; Manchur, G.; Michel, G.L.; Pastore, R.C.; et al. Summary of system protection and voltage stability. *IEEE Trans. Power Deliv.* **1995**, *10*, 631–638.

![applied sciences logo] *applied sciences*

MDPI

Article

Development of a Sequential Restoration Strategy Based on the Enhanced Dijkstra Algorithm for Korean Power Systems

Bokyung Goo, Solyoung Jung and Jin Hur *

Departement of Energy Grid, Sangmyung University, Seoul 03016, Korea; 201632070@sangmyung.kr (B.G.); 201531114@sangmyung.kr (S.J.)
* Correspondence: jinhur@smu.ac.kr; Tel.: +82-2-781-7576; Fax: +82-2-2287-0072

Academic Editors: Josep M. Guerrero and Amjad Anvari-Moghaddam
Received: 3 October 2016; Accepted: 12 December 2016; Published: 15 December 2016

Abstract: When a blackout occurs, it is important to reduce the time for power system restoration to minimize damage. For fast restoration, it is important to reduce taking time for the selection of generators, transmission lines and transformers. In addition, it is essential that a determination of a generator start-up sequence (GSS) be made to restore the power system. In this paper, we propose the optimal selection of black start units through the generator start-up sequence (GSS) to minimize the restoration time using generator characteristic data and the enhanced Dijkstra algorithm. For each restoration step, the sequence selected for the next start unit is recalculated to reflect the system conditions. The proposed method is verified by the empirical Korean power systems.

Keywords: restoration; black start service; power system restoration; generator start-up sequence

1. Introduction

The current bulk power systems require a high level of reliability. To improve the reliability of a power grid, system operators analyze N-1 contingencies, monitor the system margins and develop new technologies such as High Voltage Direct Current (HVDC) and Flexible AC Transmission Systems (FACTS) which have been introduced. Nevertheless, there is the possibility of total and partial blackouts. According to recent blackouts [1,2], large blackouts can be described as cascading outages, as shown Figure 1. Cascading outages occur sequentially, and they are caused by an initial disturbance. The initial disturbance includes natural disasters, unexpected accidents, misoperation of a facility and imbalanced power systems. It causes sequential trips of facilities such as transmission lines, transformers and generators. Sequential outages are propagated until the system fails to recover due to voltage instability and thermal violations which are alleviated or drop below the operating limits. Eventually, a partial or total blackout occurs. Once the blackouts have occurred, the appropriate system restoration should be performed.

Figure 1. Typical blackout procedure.

Most blackouts are partial and can be restored with the support of neighboring regions [3,4]. On the other hand, if total blackouts occur, neighboring regions may be not able to assist the system, especially for isolated power systems such as Korean power systems. For isolated systems, a reliable restoration method is more essential than interconnected power systems. When a total blackout occurs in isolated systems, the whole system is usually divided into several subsystems to restore the system as quickly as possible [5], and subsystems are restored in parallel. After the restoration of each subsystem, all subsystems are synchronized. Each subsystem must have at least one black start unit and should be balanced between generation and load [6]. Additionally, each subsystem must satisfy the following reliability criteria: real and reactive power balance, thermal constraints on transmission lines, being sustained over voltages during early restoration, and maintaining a steady state and transient stability during restoration.

Figure 2 shows the general black start procedure [7–9]. First, when a blackout occurs, the status of the power systems is checked through an alarm or communication. In this step, the size and status of the blackout are evaluated. Afterwards, the generators in the systems are identified using their location and capacity and the proximity to the grid. Based on the confirmed information, the sequence of the generators is established [10,11]. According to the sequence, the generators in the systems are started. While monitoring the power systems and identifying the connectivity of the system, the path for restoration is searched. The transmission lines and transformers on the paths are re-energized, and the load is restored depending on the system balance [12–14]. During the restoration stage, three limitations are needed to maintain the range: the voltage of the bus, the overload rate of the transmission lines and the frequency of the grid.

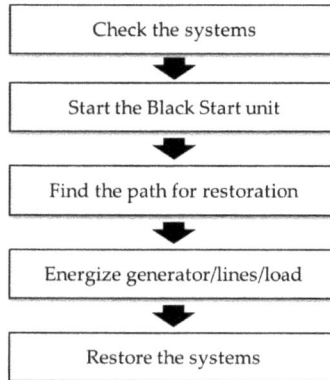

Figure 2. The general procedure of the black start service.

The restoration sequence of non-black start units, transmission lines, and transformers is different according to the objective function of each system's restoration methodology. Many domestic and foreign studies have been performed, and each Independent System Operator (ISO) has a different restoration methodology [15,16]. After a complete blackout occurs, the Australian Energy Market Operator (AEMO) in Australia aims to supply power to major generators in 90 minutes and to restore 40% of the system peak load in four hours [17–19]. The Pennsylvania New Jersey Maryland (PJM) ISO in USA intends to restore 80% of the whole system load in 16 hours and to implement the system restoration with the priority of system stability [20,21]. The Electric Reliability Council of Texas (ERCOT) ISO in USA aims to restore the whole system within the shortest amount of time [22]. At the same time, the system load should not exceed 5% of the total power in every single step. Also, when a blackout occurs, the status of the system is reported to the ERCOT ISO by the transmission system operators, and the system operators carry out the restoration procedure using the specified black start units. The system is divided into small islands and all islands are synchronized after completing the restoration process of each island. Many studies related to the methodology of restoring the islands have studied the ERCOT ISO [23]. On the other hand, the Korean Power Exchange (KPX) in Korea divides the whole system into subsystems and restores the black start units and pre-assigned paths in each subsystem first. Afterwards, there is no specific restoration methodology.

It is important to perform the restoration process considering the system condition and understanding the characteristics of the blackout. The Electrical Research Institute (EPRI) proposed a generic restoration milestones (GRM) technique to restore the system depending on various system conditions [24,25]. In the GRM, there are six milestones and system operators can select a series of milestones to restore the system depending on system conditions. The Power System Engineering Research Center (PSERC) suggested restoration procedures based on optimization techniques [6]. During the power system restoration process, the optimization technique Mixed Integer Linear Programming (MILP) is used to determine the generators' start-up sequence [26,27]. In this paper, we propose a sequential restoration strategy based on the enhanced Dijkstra algorithm for Korean power systems. In the methodology, we determine the generator start-up sequence (GSS) using generator characteristic data and an enhanced Dijkstra algorithm. To determine the GSS, we compare the characteristic data of generators, including the cranking power, start-up time, and ramping rate. Additionally, we create an adjacency matrix and consider the charging current as a weighting factor to establish power grids quickly [28]. The proposed method is verified by Korean power systems.

2. Enhanced Dijkstra Algorithm

In order to minimize damage after blackouts occur, restoration should be fulfilled reliably and quickly. For fast restoration, it is important to reduce the time taken for the selection of generators, and the time taken for the transmission lines and transformers should be considered.

First, to restore the generators, the connectivity between two different buses should be verified using an adjacent matrix [25,29]. We can find the path from the energized block which is already restored to a generator that can be restored. The adjacent matrix $\mathbf{A}(k)$ based on the transformation of the connection matrix can be presented as follows [12]:

$$\mathbf{A}(k) = \left[line_{ij}^k \right] = \begin{cases} 1 & i = j \text{ or } i \text{ \& } j \text{ are connected directly} \\ 0 & i \text{ \& } j \text{ are not connected directly} \end{cases} \tag{1}$$

where $line_{ij}$ is a line between bus i and bus j. After generating the adjacent matrix, we utilize a Dijkstra algorithm to find the shortest path to the generator. At the same time, we consider the charging current of each transmission line as a weighting factor to avoid excessive charging currents.

$$\mathbf{B}(k) = \begin{cases} 0 & i \text{ and } j \in \Omega_{E(S)} \\ a \text{ charging current of } line_{ij} & i \text{ or } j \notin \Omega_{E(S)} \\ a \text{ large number } \rho & line_{ij} \text{ is a transformer and } i \text{ or } j \notin \Omega_{E(S)} \end{cases} \tag{2}$$

where $\Omega_{E(S)}$ is the energized block that is already restored. If the number of transformers in a path is increased, the likelihood of ferroresonance may increase. However, transformers usually have small charging currents. Therefore, they should be assigned a large number of ρ as weighting factors. In this paper, we use the R program to generate the adjacent matrix.

For example, in Figure 3, if we find the paths from bus 2 to 16, we can consider two different paths. The one that is marked with the red line is the shortest path which does not consider the weighting factor, while the other that is marked with the blue dashed line is the path with the smallest weighting. Even though the red line is shorter, we will select the blue dashed line. The blue and yellow balls illustrated in Figure 3 represent the buses with bus numbers and generators respectively.

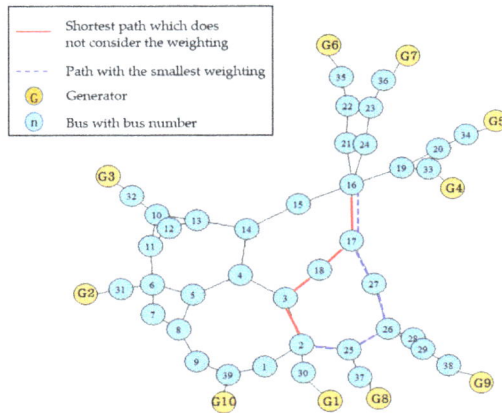

Figure 3. The example of applying the enhanced Dijkstra algorithm.

3. Sequential Restoration Strategy

3.1. Problem Formulation

The object function is used to minimize the total time taken to restore generators [30]. It should select the generator that can be restored in the shortest time. The object function can be denoted with the following equation:

$$min \sum_{i=2}^{N} \Delta t_{\text{GEN}_{i+1}} - t_{\text{GEN}_i} \tag{3}$$

where N is the total number of generators; t_{GEN_i} is the time to re-start the ith generator at the ith step; and $\Delta t_{\text{GEN}_{i+1}} - t_{\text{GEN}_i}$ means the time taken to start the $i+1$th generator from the ith generator. The object function can be formulated as the minimization of the restoration time. While restoring the system, we must satisfy the following constraints.

- Solving power flow equations

$$PowerFlow(P_{\text{GEN}}, P_{\text{LO}}, Q_{\text{GEN}}, Q_{\text{LO}}) = 0 \tag{4}$$

- No violation of generation, transmission and voltage limits

$$V_{\text{MIN\_BU}} \leq V_{\text{BU}} \leq V_{\text{MAX\_BU}} \tag{5}$$

$$P_{\text{MIN\_LO}} \leq P_{\text{LO}} \leq P_{\text{MAX\_LO}} \tag{6}$$

$$P_{\text{MIN\_GEN}} \leq P_{\text{GEN}} \leq P_{\text{MAX\_GEN}} \tag{7}$$

The second equation means the system balance P_{GEN}, P_{LO} are the active power of the generator and load, and Q_{GEN}, Q_{LO} are the reactive power of the generator and load, respectively. Equations (5)–(7) represent the need to maintain a range of voltage and power; V_{BU} is the voltage of the bus. $V_{\text{MIN\_BU}}$, $V_{\text{MAX\_BU}}$ denote the minimum and maximum, respectively. P_{LO} means the power of the load, and $P_{\text{MIN\_LO}}$ and $P_{\text{MAX\_LO}}$ are the minimum and maximum, respectively. In the same way, Equation (5) is the range of the generator.

The criteria for selecting the next generator are required as in the following equations: $\sum P_{\text{gen}}^i$ is the amount of output in MW from the ith generator; P_{start}^{i+1} is the power needed to start the $i+1$th generator. When the generation amount of the system is large enough to supply the cranking power of the generator to be committed, the sequential process of the restoration is performed. If the generation amount of the system is insufficient, it will take some time to meet the cranking power. In summary, we can select the next generator when the total amount of power to the ith is greater than the cranking power of the $i+1$th generator.

$$\sum P_{\text{gen}}^i - P_{\text{start}}^{i+1} \geq 0 \tag{8}$$

To restore the generator in a short amount of time, the generator must calculate the time required for the re-energizing. The estimation of the time for cranking the generator is performed in each step. The time can be formulated as the following equation. The generator that takes the shortest time to start is selected as the next generator by comparing the time calculated as shown below:

$$t_{\text{GEN}_i} = ST_{\text{GEN}}^i + ETL_i^{i-1} + ETr_i^{i-1} \tag{9}$$

where ST_{GEN}^i is the start-up time for the ith generator; ETL_i^{i-1} is the energizing time of the transmission lines to restore the path from $i-1$th to ith; ETr_i^{i-1} is the restoration time of the transformers to the ith generator; ST_{GEN}^j may be described in the equation and Figure 4 as follows.

$$ST_{\text{GEN}} = T_{\text{start}} + \frac{P_{\text{start}}}{Rr} \tag{10}$$

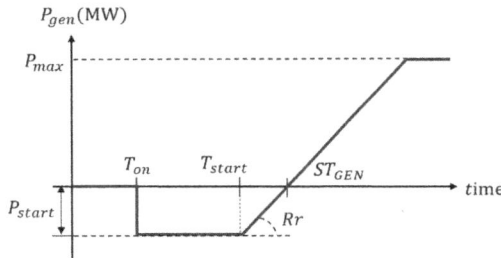

Figure 4. Generator capability curve.

When supplying power from the grid at time T_{on}, the generator starts power generation after the time required for cranking the generator. In the case of a black start unit that can be started without any external power from the power system, the T_{on} value is zero. Once the generator starts power generation at time T_{start}, it generates the power up to the maximum power according to its ramp rate, Rr.

3.2. Optimal Restoration Approach

We will propose a new sequential restoration strategy, as shown in Figure 5. First, if a blackout occurs, the restoration process is started. We set the recovery time $t = 0$ and the restoration step $i = 1$. In the next stage, the black start unit is started, and we confirm the generator that can be restored in the next step. If all generators are restored, the flow is terminated. Otherwise, the flow is continued to determine the generator start-up sequence (GSS), and the start-up time of each generator to be restored is calculated. At this time, the time taken to re-energize the transmission lines and transformers is calculated. To consider this, we search for the shortest path to the generator by considering the charging currents as a weighting factor. The time taken to restore the transmission lines and transformers is calculated by counting the amount of equipment in the path and multiplying it by a certain value. Afterwards, we compute the time to re-start the generator in accordance with the generator characteristic data. By adding the respective times that are calculated, the generator with the shortest restoration time is selected for the next unit. During the process of restoration, the constraints expressed in Equations (5)–(7) must be satisfied. Unless the constraints are satisfied, the corrective actions such as power generation or load adjustment should be performed. The restoration of the next unit is started, and the load is restored in order to maintain the system balance and stability. At this time, the sum of the time to re-energize transmission lines and transformers and the time until the generator can supply the power is added to the recovery time. In the same way, the flow of the GSS is repeated until the available generators exist in the systems.

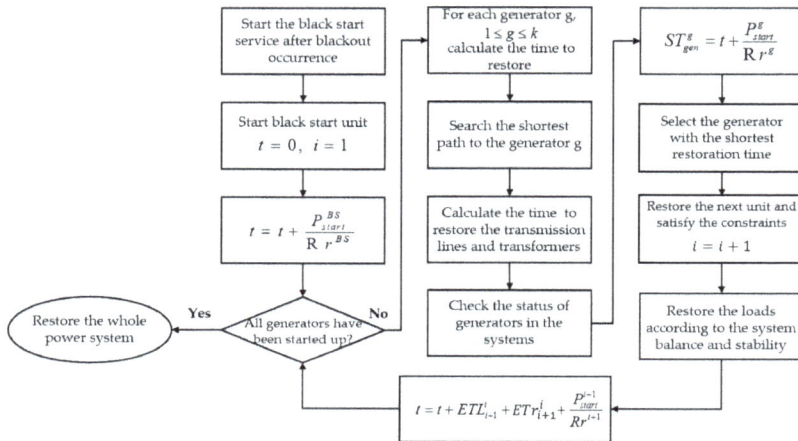

Figure 5. Flow chart of generator start-up sequence.

4. Case Study: Eastern Regions of South Korea

In Korea, when a total blackout occurs, the whole system is divided into seven regions to restore the whole system (Figure 6). Each subsystem has at least one black start unit and pre-assigned paths. Black start units are usually hydroelectric power or pumped storage power. If a blackout occurs, the black start unit and pre-assigned paths are restored first. After the black start unit and pre-assigned paths are restored, the rest of the generators are ordered to be restored.

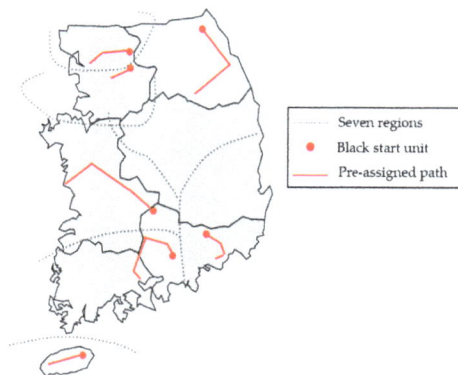

Figure 6. Korean power systems.

In this section, we assume that a total blackout has occurred, and we choose one subsystem, which is the eastern regions, among the subsystems to restore it. The black start unit and pre-assigned paths are restored first, and the rest of the generators in the eastern regions are ordered to be restored. To find the restoration sequence of the rest of the generators, the information on the generators and the charging currents of the transmission lines are used. Table 1 shows the information about the generators in the eastern regions. It does not display the exact name of the generators due to confidentiality. G6 is a black start unit, and its capacity is 280 MW. The rest of the 19 generators are non-black start units. In the table, there is information about 20 generators, such as the capability and ramping rate in MW/h, to calculate the start-up time for the next units.

Table 1. Information about the generators in the eastern regions.

Generator	*Capacity* (MW)	*Rr* (MW/h)	P_{start} (MW)	T_{start} (h)
G1	183	24.5	0.18	0.55
G2	186	24.5	0.14	1.05
G3	830	28.4	6.72	0.70
G4	150	1.5	4.14	1.70
G5	200	1.5	7.15	1.88
G6	280	100.0	0.42	0.05
G7	260	100.0	0.45	0.07
G8	255	100.0	0.50	0.09
G9	240	100.0	0.55	0.10
G10	100	22.5	0.05	0.30
G11	100	22.5	0.04	0.35
G12	35	15.0	0.00	0.08
G13	37	15.0	0.00	0.07
G14	282	18.0	0.00	0.20
G15	29	18.0	0.00	0.15
G16	30	18.0	0.00	0.18
G17	32	18.0	0.00	0.15
G18	210	1.0	11.88	7.20
G19	200	1.0	12.04	7.00
G20	180	24.5	0.15	0.89

In order to determine the GSS, the time taken to energize each generator is calculated. The generator that takes the shortest time to start is selected as the next generator. When the time is calculated, the restoration time of the transmission lines to the generator and the restoration time of the transformers to the generator are considered together, and the charging current of the transmission lines and transformers is used. Figure 7 illustrates the adjacency matrix of the eastern regions. The bus numbers are re-assigned to generate the adjacency matrix. Using this matrix, we can consider the shortest path to the non-black start unit.

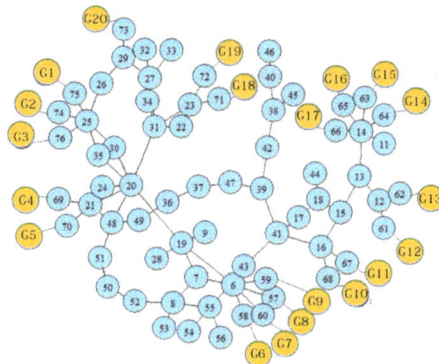

Figure 7. Adjacency matrix of the eastern regions.

Table 2 shows an example of the GSS in the eastern regions of South Korea. ST_{GEN} means the start-up time for each generator, and $ETLr$ is the energizing time of the transmission lines and transformers for restoring the generator. Total means the total time taken to restore the generator, including the time taken to restore the transmission lines and transformers. The highlighted cells are the generators that are selected as the next units in each step. In the eastern regions, the pre-assigned paths are G6 (Black start unit) $\rightarrow 58 \rightarrow 6 \rightarrow 19 \rightarrow 20 \rightarrow 30 \rightarrow 25 \rightarrow 26 \rightarrow 29 \rightarrow 73 \rightarrow$ G20, and G6 is started as the black start unit in the first step. In the second step, G20 is started as well. Afterwards, we calculate the total times for all available generators and compare them. In the third step, G7, which has the shortest total time, is restored. At the end of the third step, the total times of all generators are compared again, and in the next step, G8 is restored. G7–G9 are sequentially restored, since these are located near each other and are the same type. In the same way, the restoration sequence is repeated until the available generators exist in the systems.

Table 2. Example of the generator start-up sequence of the eastern regions.

Gen	STEP 3			STEP 4			STEP 5			STEP 6		
	ST_{GEN}	$ETLr$	Total	ST_{GEN}	$ETLr$	Total	ST_{GEN}	$ETLr$	Total	ST_{GEN}	$ETLr$	Total
G1	0.55	0.17	0.72	0.55	0.17	0.72	0.55	0.17	0.72	0.18	0.17	0.35
G2	1.05	0.17	1.22	1.05	0.17	1.22	1.05	0.17	1.22	**0.14**	**0.17**	**0.31**
G3	0.70	0.17	0.87	0.70	0.17	0.87	0.70	0.17	0.87	6.72	0.17	6.89
G4	1.70	0.08	1.78	1.70	0.08	1.78	1.70	0.08	1.78	4.14	0.08	4.22
G5	1.88	0.08	1.96	1.88	0.08	1.96	1.88	0.08	1.96	7.15	0.08	7.23
G7	**0.08**	**0.00**	**0.08**	-	-	-	-	-	-	-	-	-
G8	0.10	0.00	0.10	**0.10**	**0.00**	**0.10**	-	-	-	-	-	-
G9	0.11	0.00	0.11	0.11	0.00	0.11	**0.11**	**0.00**	**0.11**	-	-	-
G10	0.30	0.67	0.97	0.30	0.67	0.97	0.30	0.67	0.97	0.05	0.67	0.72
G11	0.35	0.67	1.02	0.35	0.67	1.02	0.35	0.67	1.02	0.04	0.67	0.71
G12	0.09	0.92	1.00	0.09	0.92	1.00	0.09	0.92	1.00	0.00	0.92	0.92
G13	0.08	0.92	0.99	0.08	0.92	0.99	0.08	0.92	0.99	0.00	0.92	0.92
G14	0.20	0.92	1.12	0.20	0.92	1.12	0.20	0.92	1.12	0.00	0.92	0.92
G15	0.16	0.92	1.08	0.16	0.92	1.08	0.16	0.92	1.08	0.00	0.92	0.92
G16	0.19	0.92	1.11	0.19	0.92	1.11	0.19	0.92	1.11	0.00	0.92	0.92

Table 3 shows the results of the simulation. According to the strategy, the generator start-up sequence (GSS) is determined using generator characteristic data and the enhanced Dijkstra algorithm. All the generators in the system are restored sequentially, and at each restoration step, the system conditions are reflected to calculate time taken. Additionally, in order to maintain the system balance and stability, sufficient load is restored during the restoration, and the total load restored in the proposed restoration strategy of the eastern regions is 2728.80 MW.

Table 3. The results of the simulation: The eastern regions.

Step	Restored Generators	Restoration Path
1	G6	-
2	G20	$58 \rightarrow 6 \rightarrow 19 \rightarrow 20 \rightarrow 30 \rightarrow 25 \rightarrow 26 \rightarrow 29 \rightarrow 73$
3	G7	60
4	G8	57
5	G9	59
6	G2	74
7	G1	75
8	G3	76
9	G11	$43 \rightarrow 41 \rightarrow 17 \rightarrow 18 \rightarrow 15 \rightarrow 16 \rightarrow 67$
10	G10	68
11	G14	$13 \rightarrow 14 \rightarrow 64$
12	G15	63
13	G16	65
14	G17	66
15	G12	$12 \rightarrow 61$
16	G13	62
17	G4	$21 \rightarrow 69$
18	G5	70
19	G18	$27 \rightarrow 34 \rightarrow 31 \rightarrow 23 \rightarrow 71$
20	G19	72

Table 4 shows the comparison of the generator start-up sequence between the proposed method and the existing method using mixed integer programming (MIP). The GSS is determined through the MIP. The existing method only considers the restoration of the maximum system capacity. The case studies from both methods are shown in the table below.

Table 4. Comparison of generator start-up sequence.

Step	Proposed Method	Existing Method
1	G6	G6
2	G20	G20
3	G7	G7
4	G8	G8
5	G9	G9
6	G2	G1
7	G1	G2
8	G3	G3
9	G11	G18
10	G10	G19
11	G14	G4
12	G15	G5
13	G16	G10
14	G17	G11
15	G12	G12
16	G13	G13
17	G4	G14
18	G5	G15
19	G18	G16
20	G19	G17

As the proposed method aims to restore the system within the shortest amount of time, the generators that can be activated in a very short time are restored preferentially. On the other hand, the existing method intends to restore the system with the maximum capacity restoration.

The time to commit all generators is 36 min for the proposed method and 55 min for the existing method. Also, the total time is 657 min using the proposed method and 643 min using the existing method when all generators reach the maximum power-generating outputs. In the proposed method, the recovery times of transmission lines and transformers are included in the restoration time; on the other hand, the time using the existing method only takes into account the start-up time of the generators.

5. Conclusions

In Korean systems, there are the pre-assigned paths and black start units in each subsystem to restore the systems. However, there is no specific restoration methodology afterwards. Therefore, it is essential to develop a restoration strategy for determining the generator start-up sequence (GSS) and the paths that can re-energize the generators according to the GSS. Also, this restoration strategy should be performed within the shortest time to minimize the economic and social loss. In this paper, we propose a sequential and systematic restoration strategy that can minimize the generator restoration time and considers the characteristics of the Korean power system.

In this paper, we propose a sequential restoration strategy based on an enhanced Dijkstra algorithm for Korean power systems. The new methodology is intended to minimize the time taken to restore the generators that can be restored. At this time, some loads are restored to maintain the system voltage and the stability of generating units. In the methodology, we developed a strategy for the determination of the generator start-up sequence in order to restore the system in the shortest time, an adjacency matrix was created, and the charging current of the path was considered as the weighting factor. The proposed method was verified by an empirical system (in the eastern regions of South Korea). In the future, we will apply the proposed methodology to other regions.

Acknowledgments: This research was supported by Basic Science Research Program through the National Research Foundation of Korea (NRF) funded by the Ministry of Science, ICT & Future Planning (NRF-2015R1C1A1A02037716)

Author Contributions: Jin Hur conceived and designed the overall research; Bokyung Goo developed the sequential restoration strategy and conducted the experimental simulation; Solyoung Jung implemented the enhanced Dijkstra algorithm and applied the enhanced algorithm to Korean power systems; Jin Hur, Bokyung Goo and Solyoung Jung wrote the paper; and Jin Hur guided the research direction and supervised the entire research process.

Conflicts of Interest: The authors declare no conflict of interest.

References

1. ENTSO-E Project Group Turkey. Report on Blackout in Turkey on 31st March 2015. Available online: https://www.entsoe.eu/Documents/SOC%20documents/Regional_Groups_Continental_Europe/20150921_Black_Out_Report_v10_w.pdf (accessed on 16 November 2016).
2. Report on the Grid Disturbance on 30th July 2012 and Grid Disturbance on 31th July 2012. Available online: http://www.cercind.gov.in/2012/orders/Final_Report_Grid_Disturbance.pdf (accessed on 16 November 2016).
3. Adibi, M.; Clelland, P.; Fink, L.; Happ, H.; Kafka, R.; Raine, J.; Scheurer, D.; Trefny, F. Power system restoration—A task force report. *IEEE Trans. Power Syst.* **1987**, *2*, 271–277. [CrossRef]
4. Adibi, M.M.; Borkoski, J.N.; Kafka, R.J. Power system restoration—The second task force report. *IEEE Trans. Power Syst.* **1987**, *2*, 927–932. [CrossRef]
5. Park, Y.-M.; Lee, K.-H. Application of expert system to power system restoration in sub-control center. *IEEE Trans. Power Syst.* **1997**, *12*, 629–635. [CrossRef]
6. Liu, C.C.; Vittal, V.; Heydt, G.T.; Tomsovic, K.; Sun, W. *Development and Evaluation of System Restoration Strategies from a Blackout*; Power System Engineering Research Center: Tempe, AZ, USA, 2009.
7. Adibi, M.M.; Martins, N. Power system restoration dynamics issues. In Proceedings of the Power and Energy Society General Meeting-Conversion and Delivery of Electrical Energy in the 21st Century, Pittsburgh, PA, USA, 20–24 July 2008.
8. Adibi, M.M.; Fink, L.H. Power system restoration planning. *IEEE Trans. Power Syst.* **1994**, *9*, 22–28. [CrossRef]
9. Adibi, M.M.; Fink, L.H. Special considerations in power system restoration. *IEEE Trans. Power Syst.* **1992**, *7*, 1419–1424. [CrossRef]
10. Sun, W.; Liu, C.C.; Chu, R.F. Optimal generator start-up strategy for power system restoration. In Proceedings of the 15th International Conference on Intelligent System Applications to Power Systems, Curitiba, Brazil, 8–12 November 2009.
11. Sun, W.; Liu, C.C.; Zhang, L. Optimal generator Start-up strategy for bulk power system restoration. *IEEE Trans. Power Syst.* **2011**, *26*, 1357–1366. [CrossRef]
12. Fink, L.H.; Liou, K.-L.; Liu, C.-C. From generic restoration actions to specific restoration strategies. *IEEE Trans. Power Syst.* **1995**, *10*, 745–752. [CrossRef]
13. Perez-Guerrero, R.E.; Heydt, G.T. Distribution system restoration via subgradient based Lagrangian relaxation. *IEEE Trans. Power Syst.* **2008**, *23*, 1162–1169. [CrossRef]
14. Perez-Guerrero, R.F.; Heydt, G. Viewing the distribution restoration problem as the dual of the unit commitment problem. *IEEE Trans. Power Syst.* **2008**, *23*, 807–808. [CrossRef]
15. Henderson, M.; Rappold, E.; Feltes, J.; Grande-Moran, C.; Durbak, D.; Bileya, O. Addressing restoration issues for the ISO new England system. In Proceedings of the Power and Energy Society General Meeting, San Diego, CA, USA, 22–26 July 2012.
16. NERC Standard EOP-005-2, System Restoration from Black Start Resources. Available online: http://www.nerc.com/files/EOP-005-2.pdf (accessed on 16 November 2016).
17. AEMO. Independent review of system restart ancillary service process improvements. Available online: https://www.aemo.com.au/-/media/Files/PDF/Independent-Review-of-System-Restart-Ancillary-Services-Process-Improvem.ashx (accessed on 16 November 2016).
18. AEMO. Interim System Restart Standard. Available online: http://www.aemc.gov.au/getattachment/c03f9653-d44d-46c7-b408-998a22b67324/AEMO-s-Interim-System-Restart-Standard.aspx (accessed on 16 November 2016).

19. AEMO. An Introduction to Australia's National Electricity Market Wholesale Market Operation. Available online: http://www.abc.net.au/mediawatch/transcripts/1234_aemo2.pdf (accessed on 16 November 2016).
20. PJM Manual 14B: PJM Region Transmission Planning Process. Available online: www.pjm.com/~/media/documents/manuals/m14b.ashx (accessed on 16 November 2016).
21. Kafka, R.J. Review of PJM restoration practices and NERC restoration standards. In Proceedings of the Power and Energy Society General Meeting-Conversion and Delivery of Electrical Energy in the 21st Century, Pittsburgh, PA, USA, 20–24 July 2008; pp. 1–5.
22. ERCOT Nodal Operating Guides, Section 8, Attachment A, Detailed Black Start Information. Available online: http://www.ercot.com/content/mktrules/guides/noperating/2013/1007/October_7, _2013_Nodal_Operating_Guides.pdf (accessed on 16 November 2016).
23. Saraf, N.; Mclntyre, K.; Dumas, J.; Santoso, S. The annual black start service selection analysis of ERCOT grid. *IEEE Trans. Power Syst.* **2009**, *24*, 1867–1874. [CrossRef]
24. Hou, Y.; Liu, C.-C.; Sun, K.; Zhang, P.; Liu, S.; Mizumura, D. Computation of milestones for decision support during system restoration. *IEEE Trans. Power Syst.* **2011**, *26*, 1399–1409. [CrossRef]
25. Hou, Y.; Liu, C.-C.; Sun, K.; Zhang, P.; Sun, K. Constructing power system restoration strategies. In Proceedings of the International Conference on Electrical and Electronics Engineering, Bursa, Turkey, 5–8 November 2009; pp. I8–I13.
26. Lee, J.; Kim, D.; Cha, J.; Kim, Y.; Joo, S.K. Optimization-based generator start-up sequence determination method using linear transformation for power system restoration in the eastern region of Korea. In Proceedings of the International Conference on Electrical Engineering (ICEE) 2016, Okinawa, Japan, 3–7 July 2016.
27. Nagata, T.; Hatakeyama, S.; Yasouka, M.; Sasaki, H. An efficient method for power distribution system restoration based on mathematical programming and operation strategy. In Proceedings of the International Conference on Power System Technology, Perth, Western Australia, 4–7 December 2000; pp. 1545–1550.
28. Crane, A.T. Physical vulnerability of electric systems to natural disasters and sabotage. *Terrorism* **1990**, *13*, 189–190. [CrossRef]
29. Chou, Y.-T.; Liu, C.-W.; Wang, Y.J.; Wu, C.C.; Lin, C.C. Development of a black start decision supporting system for isolated power systems. *IEEE Trans. Power Syst.* **2013**, *28*, 2202–2210. [CrossRef]
30. Ancona, J.J. A framework for power system restoration following a major power failure. *IEEE Trans. Power Syst.* **1995**, *10*, 1480–1485. [CrossRef]

applied sciences

MDPI

Review

Virtual Inertia: Current Trends and Future Directions

Ujjwol Tamrakar [1], Dipesh Shrestha [1], Manisha Maharjan [1], Bishnu P. Bhattarai [2], Timothy M. Hansen [1] and Reinaldo Tonkoski [1,*]

[1] Department of Electrical Engineering and Computer Science, South Dakota State University, Brookings, SD 57007, USA; ujjwol.tamrakar@jacks.sdstate.edu (U.T.); dipesh.shrestha@jacks.sdstate.edu (D.S.); manisha.maharjan@jacks.sdstate.edu (M.M.); timothy.hansen@sdstate.edu (T.M.H.)
[2] Department of Power and Energy Systems, Idaho National Laboratory, Idaho Falls, ID 83415, USA; bishnu.bhattarai@inl.gov
* Correspondence: reinaldo.tonkoski@sdstate.edu; Tel.: +1-605-688-6298

Received: 30 April 2017; Accepted: 19 June 2017; Published: 26 June 2017

Abstract: The modern power system is progressing from a synchronous machine-based system towards an inverter-dominated system, with large-scale penetration of renewable energy sources (RESs) like wind and photovoltaics. RES units today represent a major share of the generation, and the traditional approach of integrating them as grid following units can lead to frequency instability. Many researchers have pointed towards using inverters with virtual inertia control algorithms so that they appear as synchronous generators to the grid, maintaining and enhancing system stability. This paper presents a literature review of the current state-of-the-art of virtual inertia implementation techniques, and explores potential research directions and challenges. The major virtual inertia topologies are compared and classified. Through literature review and simulations of some selected topologies it has been shown that similar inertial response can be achieved by relating the parameters of these topologies through time constants and inertia constants, although the exact frequency dynamics may vary slightly. The suitability of a topology depends on system control architecture and desired level of detail in replication of the dynamics of synchronous generators. A discussion on the challenges and research directions points out several research needs, especially for systems level integration of virtual inertia systems.

Keywords: frequency stability; microgrid control; renewable energy; virtual inertia

1. Introduction

The demand for clean energy in the modern power system is on the rise, driven by factors such as fuel prices, laws, and regulations. Renewable energy sources (RESs) like photovoltaic (PV) and wind energy are now gradually starting to dominate the energy generation mix, replacing traditional generation sources, such as coal and nuclear [1,2]. The popularity of distributed PV plants further escalates the penetration of renewables in the modern power system. The global installation of wind and PV generation exceeded 400 GW and 200 GW, respectively, by the end of 2015 [3]. Countries like Ireland and Germany already have annual RES penetrations of more than 20% [4]. In Denmark, wind power alone has the capacity to meet 40% of the country's instantaneous electricity demand, which is the highest among all the countries. The rapid development of RES is causing the modern power grid to gravitate towards an inverter-dominated system from a rotational generator-dominated system, as illustrated in Figure 1. PV systems and most modern wind turbines are interfaced through inverters. Although this is advantageous from the point-of-view of harvesting RES, the inverter-based generation does not provide any mechanical inertial response, and hence compromises frequency stability [4–6].

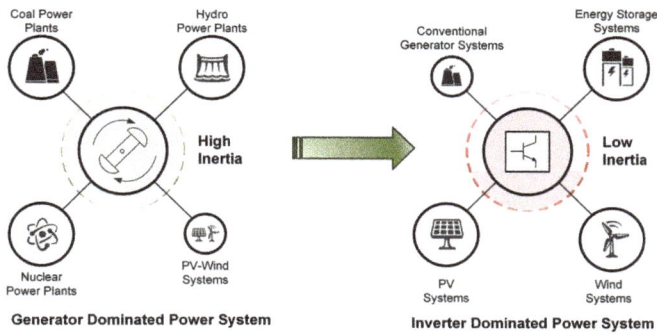

Figure 1. Evolution towards an inverter dominated power system.

Recent reports and studies have shown frequency stability to be a matter of significant concern due to lack of inertial response from RESs. The independent system operator, Electricity Reliability Council of Texas (ERCOT) has reported a continuous decline in the inertial response of its system and recommends additional inertial response [7,8]. Figure 2 illustrates the change in frequency in the ERCOT interconnection for two time periods for the same amount of generation loss. The change in frequency per generation loss is increasing yearly, and this trend is highly correlated with increased RES penetration over the same time-period. Similarly, the European Network of Transmission System Operators for Electricity (ENTSO-E) has reported increased frequency violations in the Nordic grid correlated with increased RES penetration [9]. As a consequence, inertial response from wind turbines is now mandatory in many countries [10,11] and the trend is extending towards PV plants as well. Accordingly, there is a strong practical relevance to research on virtual inertia systems which was of an academic nature in the past.

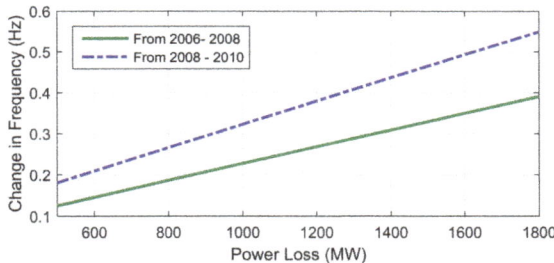

Figure 2. Increase in frequency changes in Electricity Reliability Council of Texas (ERCOT) connection due to generation loss [7].

In order to maintain the power generation and load balance, various control actions are implemented in a power system over multiple time-frames as illustrated in Figure 3. The governor response is the primary control action which takes place within the first few seconds (typically 10–30 s) of a frequency event and aims at reducing the frequency deviation. The automatic generation control is the secondary control action which takes place within minutes (typically 10–30 min) and restores the system frequency back to the nominal value. The tertiary control action is the reserve deployment when actions are taken to get the resources in place to handle present or future disturbances in the system. Whenever there is an imbalance between the generation and consumption in a power system, the generators cannot respond instantaneously to balance the system. The kinetic energy stored in the rotors is responsible for counteracting this imbalance through inertial response until the primary frequency control has been activated. As conventional generators are displaced by RESs, the inertial response also decreases. This leads to an increased rate-of-change-of-frequency (ROCOF), and a low frequency nadir (minimum frequency point) in a very short time. The primary frequency control cannot respond within the small time frame (typically less than 10 s) to arrest the system frequency change. This period is highlighted as section AB in Figure 3. It is clear from the figure that in systems with lower inertia, the frequency nadir is considerably lower along with a high ROCOF. Such situations can lead to tripping of frequency relays (causing under-frequency load shedding (UFLS)) and, in the worst case, may lead to cascaded outages [12,13]. The solution to such scenarios is to add virtual inertia in the system. The basic requirements of a virtual inertia system is that it has to operate in a very short time interval (typically less than 10 s) and in autonomous fashion. Deployed appropriately, virtual inertia systems would enhance system stability and enable greater penetration of RESs.

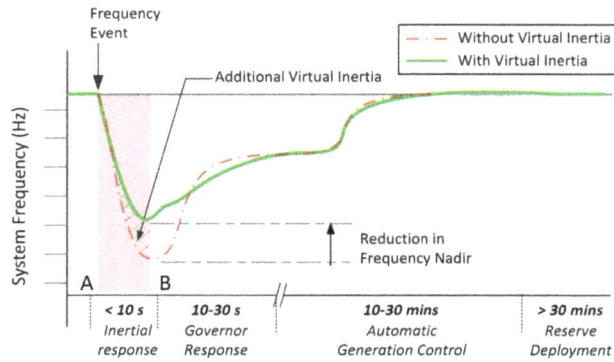

Figure 3. Multiple time-frame frequency response in a power system following a frequency event.

This paper presents a literature review of the various topologies used for virtual inertia implementation. The major topologies and the consequent improvements in these topologies are reviewed through a literature search followed by a restudy through simulations. The problem of large frequency variations due to high penetration of RESs are introduced first in Section 2. The "first generation" of virtual inertia systems are introduced next in Section 3. The topologies and control algorithms to effectively emulate inertia of synchronous generators (SGs) through power electronic based converters are discussed. After a literature review of the virtual inertia topologies, three main topologies are compared and evaluated in a common benchmark in Section 4. The "second generation" of virtual inertia systems is then reviewed in Section 5. The optimization of these systems in terms of dynamic performance and energy usage is discussed. Finally, a review of the challenges involved with integrating virtual inertia systems into the existing power system and some future research directions are discussed in Section 6. Section 7 discusses the conclusions of the paper.

2. Frequency Variations in Weak Power Systems with High Penetration of RES

Microgrids have been identified as the best option to integrate distributed generation (DG) units in terms of flexibility and reliability [14–16]. The microgrids can be operated in three possible modes: grid-connected, islanded, or isolated. A microgrid is said to have been islanded when a microgrid that is grid-connected disconnects from the grid, either in a planned fashion or due to a fault/disturbance in the main grid. In the isolated mode of operation, the microgrid is designed such that it is never connected to the grid. Regardless, these microgrid systems represent weak power systems and the high penetration of inertia-less PV and wind energy systems has a severe effect on the frequency stability. The rapid changes in the generation can cause frequency variations in the system that are outside standard limits and compromise the stability of the system.

Figure 4 shows the recommended standard frequency range for grid-connected and isolated/islanded microgrids. In the grid-connected mode, the frequency is controlled by the main grid and the frequency deviations are relatively small. However, this scenario is slowly changing with increased integration of large-scale inertia-less generation. The Institute of Electrical and Electronic Engineers (IEEE) recommends a tight frequency operating standard of ± 0.036 Hz for grid-connected systems. The North American Reliability Corporation (NERC) recommends triggering the first level of UFLS when the system frequency drops below 59.3 Hz (for a nominal frequency of 60 Hz for the US power grid). The activation of UFLS is the last automated reliability measure to counteract frequency drop and re-balance the system [17]. NERC recommended control actions include disconnecting the generator if the frequency drops below 57 Hz or rises above 61.8 Hz [18]. The European Norm EN50160 also imposes similar tight ranges for grid-connected microgrid systems [19]. There are no specific standards defined for frequency limits for isolated microgrid systems. This is highly dependent on the generation and the load mix in a particular microgrid system. From a generator point-of-view, frequency standards like the ISO 8528-5 standard [20] can provide a guideline for the frequency limits. With the small amount of SGs in isolated microgrids, the frequency excursions and ROCOF are greater and the need for virtual inertia is of high importance. In such isolated microgrids, to implement virtual inertia, either dedicated energy storage systems (ESSs) can be used [21,22], or inertia can be emulated by operating PV/wind below their maximum power point (MPP) [23,24]. However, the allowable frequency nadirs and ROCOFs in the microgrids in islanded/isolated conditions may be relaxed compared to grid-connected operation. This will be especially vital for the design of virtual inertia systems for isolated microgrids as these microgrids often have limited energy resources and relaxing the frequency operating region would result in significant energy saving and reduction in power ratings of virtual inertia systems.

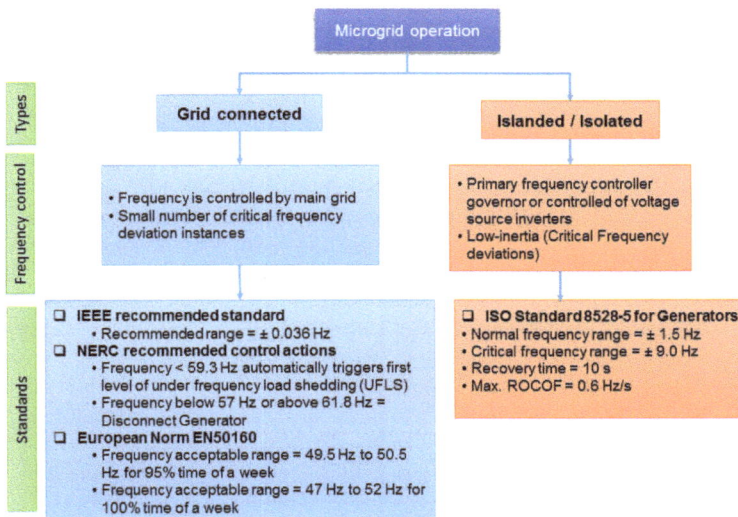

Figure 4. Frequency standards for microgrid systems [18–20].

3. First Generation: Virtual Inertia Topologies

3.1. Concept and Classification of Virtual Inertia Topologies

The frequency variation in a power system after a frequency event/disturbance can be approximated by the swing equation [25]:

$$P_{gen} - P_{load} = \frac{d(E_{K.E.})}{dt} = \frac{d(\frac{1}{2}J\omega_g^2)}{dt} \tag{1}$$

$$P_{gen} - P_{load} = J\omega_g \frac{d\omega_g}{dt} \tag{2}$$

where, P_{gen} is the generated power, P_{load} is the power demand including losses, J is the total system inertia, and ω_g is the system frequency. The inertia constant of the power system H is the kinetic energy normalized to apparent power S_g of the connected generators in the system:

$$H = \frac{J\omega_g^2}{2S_g} \tag{3}$$

Equation (2) can then be written as:

$$\frac{2H}{\omega_g}\frac{d\omega_g}{dt} = \frac{P_{gen} - P_{load}}{S_g} \tag{4}$$

Equation (4) can also be represented in terms of frequency (Hz) instead of angular frequency (rad/s) as follows:

$$\frac{2H}{f}\frac{df}{dt} = \frac{P_{gen} - P_{load}}{S_g} \tag{5}$$

where, $\frac{df}{dt}$ is the ROCOF of the system. With reduced inertia, the ROCOF of the system increases which causes larger changes in frequency of the system in the same time-frame. Thus, the system requires additional inertia as more RESs are integrated into the power system. The concept of virtual inertia implementation using power electronic converters was first developed by Beck and Hesse [26]. Many other topologies and approaches have been developed in the literature since.

Virtual inertia is a combination of control algorithms, RESs, ESSs, and power electronics that emulates the inertia of a conventional power system [13]. The concept of virtual inertia is summarized in Figure 5. The core of the system is the virtual inertia algorithm that presents the various energy sources interfaced to the grid through power electronics converters as SGs. Most modern wind turbines are operated as variable speed wind turbines and interfaced through back-to-back converters, completely decoupling the inertia from the grid. Similarly, PV systems and ESSs have a DC-DC converter and an inverter in the front-end, and do not contribute to the inertial response [4,27]. Virtual inertia systems based on current/voltage feedback from the inverter output generate appropriate gating signals to present these resources as SGs from the point-of-view of the grid [28]. Although the basic underlying concepts are similar among the various topologies in the literature, the implementation is quite varied based on the application and desired level of model sophistication. Some topologies try to mimic the exact behavior of the SGs through a detailed mathematical model that represent their dynamics. Other approaches try to simplify this by using just the swing equation to approximate the behavior of SGs, while others employ an approach which makes the DG units responsive to frequency changes in the power system. This section discusses the various topologies that have been proposed in literature. Figure 6 shows a general classification of various topologies that are available in the literature for virtual inertia implementation. Among the listed topologies, the synchronverter, the Ise lab's topology, the virtual synchronous generator (most popular in literature from each classification), and the droop control were selected for a detailed description. A brief description of the remaining topologies is also presented under Section 3.6.

Virtual Inertia Emulation

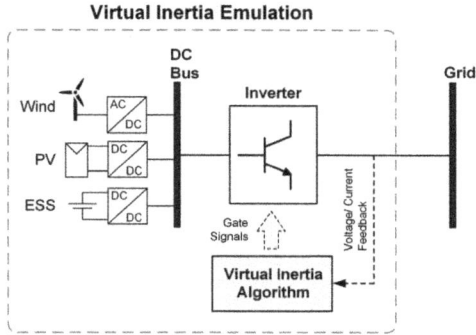

Figure 5. Concept of virtual inertia.

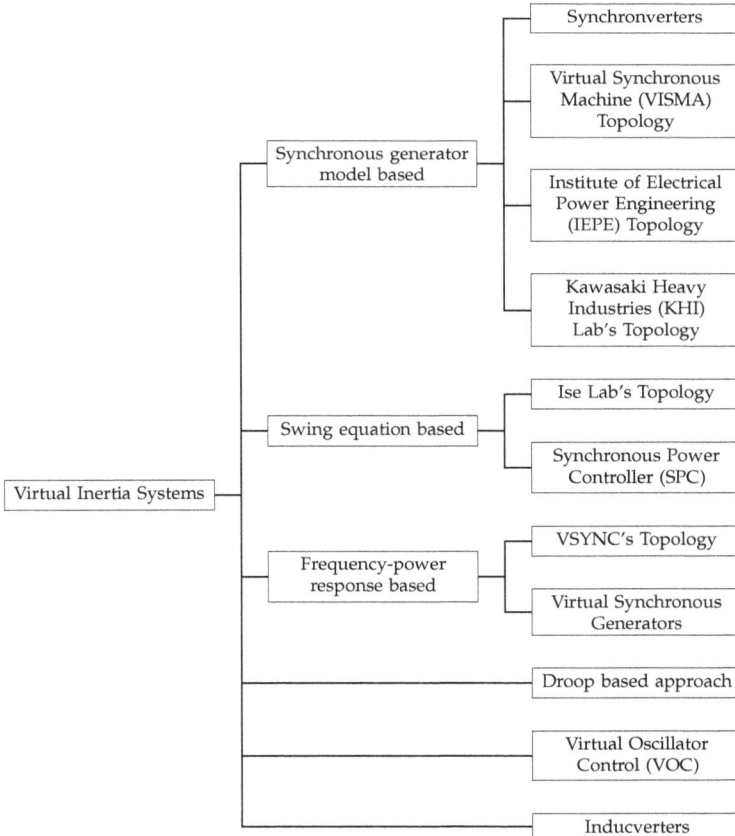

Figure 6. Classification of different topologies used for virtual inertia implementation.

3.2. Synchronverters: A Synchronous Generator Model Based Topology

Synchronverters operate the inverter-based DG units as SGs representing the same dynamics from the point-of-view of the grid [29]. This is based on the notion that such a strategy allows traditional operation of the power system to be continued without major changes in the operation structure. The topology is well developed in the literature by Q.C. Zhong [30]. A frequency drooping mechanism is used to regulate the power output from the inverter similar to how the SG regulates its power output [31]. The following basic equations are used to capture the dynamics of the SG:

$$T_e = M_f i_f < i, \widetilde{sin\theta} >$$ (6)

$$e = \dot{\theta} M_f i_f \widetilde{sin\theta}$$ (7)

$$Q = -\dot{\theta} M_f i_f < i, \widetilde{cos\theta} >$$ (8)

where, T_e is the electromagnetic torque of the synchronverter, M_f is the magnitude of the mutual inductance between the field coil and the stator coil, i_f is the field excitation current, θ is the angle between the rotor axis and one of the phases of the stator winding, e is the no load voltage generated, and Q is the generated reactive power. In Equations (6) and (8), $\langle \cdot, \cdot \rangle$ represents the standard inner product of two vectors in $\mathbb{R}^3$. The three-phase stator current, i, $\widetilde{sin\theta}$, and $\widetilde{cos\theta}$ are vectors defined as follows:

$$i = \begin{bmatrix} i_a \\ i_b \\ i_c \end{bmatrix} ; \widetilde{sin\theta} = \begin{bmatrix} sin\theta \\ sin(\theta - \frac{2\pi}{3}) \\ sin(\theta - \frac{4\pi}{3}) \end{bmatrix} ; \widetilde{cos\theta} = \begin{bmatrix} cos\theta \\ cos(\theta - \frac{2\pi}{3}) \\ cos(\theta - \frac{4\pi}{3}) \end{bmatrix}$$ (9)

Equations (6)–(8) are first discretized and then solved in each control cycle in a digital controller to generate the gating signals for the DG unit under consideration. Figure 7a shows the basic schematic of the synchronverter. The dashed box represents the control part of the synchronverter, the details of which are illustrated in Figure 7b. The inverter output current i and grid voltage v are the feedback signals utilized to solve the differential equations within the controller. Additionally, the desired moment of inertia J and damping factor D_p can be set as desired. The selection of these parameters is crucial from the point-of-view of the stability of the system as shown in [32]. The frequency and voltage loops, as indicated in Figure 7b, are used to generate the control inputs—the mechanical torque, T_m and $M_f I_f$. In the frequency loop, T_m is generated from the reference active power P^* based on the nominal angular frequency of the grid w_n. The virtual angular frequency of the synchronverter w is thus generated by this loop which is integrated to calculate the phase command θ and is used for the pulse width modulation (PWM). Similarly, in the voltage loop, the difference between the reference voltage v^* and and the amplitude of the grid voltage v is multiplied by a voltage drooping constant D_q. This is added to the error between the reference reactive power Q^* and the reactive power Q calculated using (8). The resulting signal is then passed through an integrator with gain $\frac{1}{K_v}$ to generate $M_f I_f$. The outputs of the controller are e and θ which are used for PWM generation.

The underlying equations of a synchronverter topology form an enhanced phase locked loop (PLL) or a sinusoid-locked loop, making it inherently capable of maintaining synchronism with the terminal voltage [33]. Single phase variants of the synchronverter have also been designed in [34]. The basic version of synchronverter requires a PLL to initially synchronize with the grid, however the use of PLLs in weak grids is known to be prone to instabilities [35–37]. To counteract this, self-synchronized synchronverters are introduced in [38]. The synchronverter topology has also inspired the operations of rectifiers as synchronous motors [39] which helps in obtaining inertial response from the load side of the power system. Moreover, the voltage-source based implementation means that synchronverters can be operated as grid forming units, and ideally suited for inertia emulation from DGs that are not connected with the main grid. The fact that the frequency derivative is not required for the implementation, is a major advantage as derivative terms often induce noise in the system. Although the synchronverter is able to replicate the exact dynamics of a SG, the complexity of the differential equations used can result in numerical instability. Moreover, a voltage-source based implementation means there is no inherent protection against severe grid transients, which may result in need of external protection systems for safe operation.

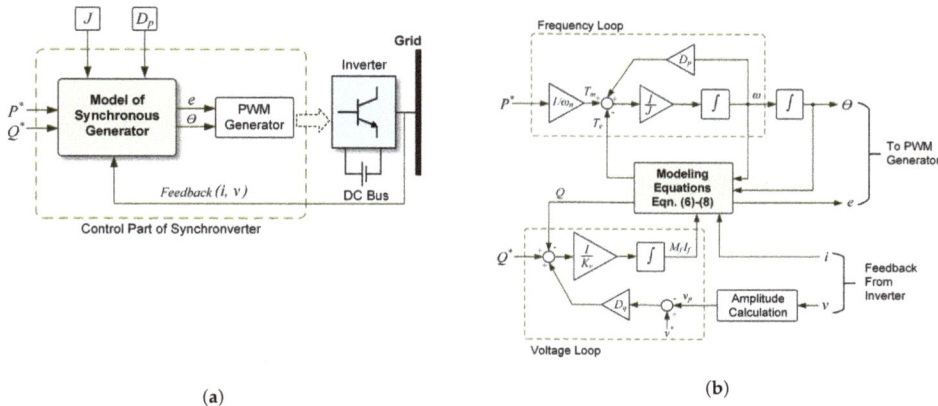

Figure 7. Synchronverter topology: (**a**) overall schematic showing operating principle; (**b**) detailed control diagram showing the modeling equations.

3.3. Ise Lab's Topology: A Swing Equation Based Topology

The topology developed by Ise lab for virtual inertia implementation is similar to the synchronverter approach described previously, but instead of using a full detailed model of the SG, the topology solves the power-frequency swing equation every control cycle to emulate inertia [40]. The schematic diagram of the topology illustrating the operation principle is shown in Figure 8a. The controller senses the inverter output current i and the voltage of the point of connection v, and computes the grid frequency ω_g and active power output of the inverter P_{out}. These two parameters are inputs to the main control algorithm block along with P_{in} which is the prime mover input power [41]. Within the control algorithm, the swing equation given by Equation (10) is solved every control cycle thus generating the phase command θ for the PWM generator. The typical swing equation of a SG is:

$$P_{in} - P_{out} = J\omega_m(\frac{d\omega_m}{dt}) + D_p\Delta\omega \tag{10}$$

$$\Delta\omega = \omega_m - \omega_g \tag{11}$$

where, P_{in}, P_{out}, ω_m, ω_g, J, and D_p are the input power (similar to the prime mover input power in a SG), the output power of the inverter, virtual angular frequency, grid/reference angular frequency, moment of inertia, and the damping factor, respectively. A model of the governor, as shown in Figure 8b, is utilized to compute the input power P_{in} based on the frequency deviation from a reference frequency ω^*. The governor is modeled as a first-order lag element with gain K and time-constant T_d. P_0 represents continuous power reference for the DG unit. The delay in the governor model leads to higher ROCOF and thus higher frequency nadirs as a consequence. The voltage reference e can be generated through $Q - v$ droop approach as described in [42,43].

Similar to the synchronverter, derivative of frequency is not needed to implement the control algorithm. This is highly beneficial as frequency derivatives are know to introduce noise in the system which makes the system difficult to control. Additionally, this topology can be used to operate DG units as grid forming units. However, problems related to numerical instability still remain, which along with improper tuning of parameters J and D_p, can lead to oscillatory system behavior [41].

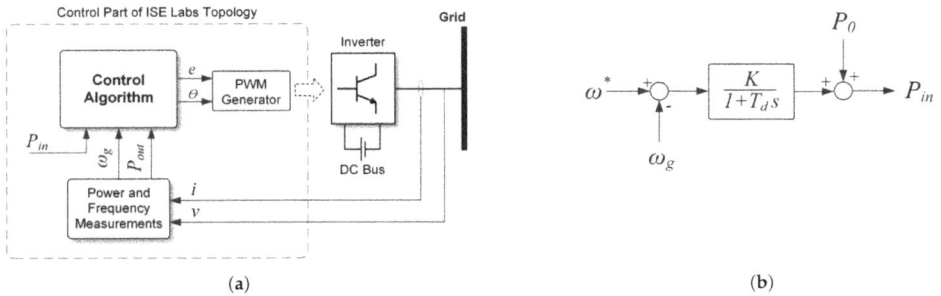

Figure 8. Ise Lab's topology: (**a**) overall schematic showing operating principle; (**b**) the governor model to compute input power.

3.4. Virtual Synchronous Generators: A Frequency-Power Response Based Topology

The main idea behind virtual synchronous generators (VSG) is to emulate the inertial response characteristics of a SG in a DG system, specifically the ability to respond to frequency changes [25,44]. This emulates the release/absorption of kinetic energy similar to that of a SG, thus presenting the DG units as a dispatchable source [45,46]. Compared to traditional droop controllers which provide only frequency regulation, the VSG approach is able to provide dynamic frequency control [21]. This dynamic control is based on the derivative of the frequency measurement and behaves similarly to inertial power release/absorption by a SG during a power imbalance. Thus, the VSG is a dispatchable current source that regulates its output based on system frequency changes. This is one of the simplest approaches to implement virtual inertia in DG systems as it does not incorporate all the detailed equations involved in a SG. However, operating multiple DG units as current sources is known to result in instability [47].

The output power of the VSG converter is controlled using Equation (12):

$$P_{VSG} = K_D \Delta \omega + K_I \frac{d\Delta \omega}{dt} \tag{12}$$

where, $\Delta \omega$ and $\frac{d\Delta \omega}{dt}$ represent the change in angular frequency and the corresponding rate-of-change. K_D and K_I represent the damping and the inertial constant, respectively. The damping constant is similar to the frequency droop and helps return the frequency to a steady-state value and reduce the frequency nadir. The inertial constant arrests the ROCOF by providing fast dynamic frequency response based on the frequency derivative. This feature is especially important in an isolated grid where the initial ROCOF can be very high, leading to unnecessary triggering of protection relays. The VSG topology is illustrated in Figure 9. A PLL is used to measure the change in system frequency and ROCOF [45]. Then, using Equation (12), the active power reference for the inverter is computed. The current references are then generated for the current controller based on this reference power. The topology illustrated here assumes a direct-quadrature (*d*-*q*) based current control approach, but any other current control techniques (as described in [48,49]) may be used. For *d*-*q* control, *d*-axis current reference can be calculated as [22]:

$$I_d^* = \frac{2}{3} \left(\frac{V_d P_{VSG} - V_q Q}{V_d^2 + V_q^2} \right) \tag{13}$$

where, V_d and V_q are the $d-axis$ and $q-axis$ components of the measured grid voltage v. The q-axis current reference I_q^* and the reactive power Q is set to zero as it is assumed that only the active power is being controlled. The current controller based on the grid current feedback generates the gate signals to drive the inverter. Thus, the inverter behaves as a current-controlled voltage source inverter [13,48].

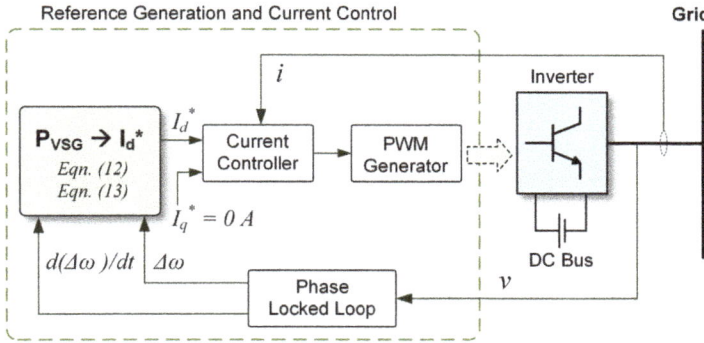

Figure 9. Virtual synchronous generator (VSG) topology.

This topology is used by the European VSYNC research group [45,50] and has demonstrated the effectiveness of inertia emulation using VSG topology through real-time simulations [51] and several field tests [52]. In [22], an experimental verification of the topology is presented for remote microgrid applications. The VSG topology has also been widely employed for virtual inertia emulation from wind systems as reported in [6,53,54]. The main drawback of this topology is that it cannot be implemented in islanded modes where the virtual inertia unit has to operate as a grid forming unit. Moreover, the system emulates inertia during frequency variations, but not in input power variations [55]. Accurate measurement of the frequency derivative through PLLs can be challenging for this kind of implementation [56,57]. The performance of PLLs can degrade and compete against each other, especially in weak grids [58,59]. PLL systems are known to show steady-state errors and instability especially in weak grids with frequency variations, harmonic distortions, and voltage sags/swells [35–37]. In [60], it was shown that the problems with instability are even more pronounced when a proportional-integral (PI) controller is used to implement the inner-current control loop of the inverter. Accordingly, a VSG requires a robust and sophisticated PLL for a successful implementation [61]. Another disadvantage of the VSG approach is that the derivative term used to compute the ROCOF makes the VSG sensitive to noise which can lead to unstable operation.

3.5. Droop-Based Approaches

The approaches described so far try to mimic or approximate the behavior of SGs to improve inertial response of inverter-dominated power systems. Different from these techniques, the frequency-droop based controllers have been developed for autonomous operation of isolated microgrid systems [62,63]. Based on the assumption that the impedance of the grid is inductive, the frequency droop is implemented as:

$$\omega_g = \omega^* - m_p(P_{out} - P_{in}) \tag{14}$$

where, ω^* is the reference frequency, ω_g is the local grid frequency, P_{in} is the reference set active power, P_{out} is the measured active power output from the DG unit, and m_p is the active power droop. Similarly, the voltage-droop is implemented as:

$$v_g = v^* - m_q(Q_{out} - Q_{in}) \tag{15}$$

where, v^* is the reference voltage, v_g is the local grid voltage, Q_{in} is the reference set reactive power, Q_{out} is the measured reactive power output from the DG unit, and m_q is the reactive power droop.

The schematic of a frequency-droop controller based on Equation (14) is shown in Figure 10. Often a low pass filter with a time constant T_f is used when measuring the output power to filter out high frequency components from the inverter [14]. In the literature [59,64,65], it has already been shown that the use of this filter makes the droop-based control approximate the behavior of virtual inertia systems. The proof was first presented by Arco et al. [59] and is repeated here for convenience.

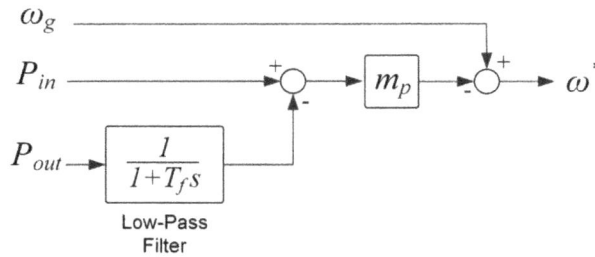

Figure 10. Schematic for frequency droop control.

Proof. Based on the schematic of Figure 10:

$$P_{out} = (1 + T_f s) \left\{ \frac{1}{m_p} (\omega_g - \omega^*) + P_{in} \right\} \tag{16}$$

Rearranging,

$$P_{in} - P_{out} = \frac{1}{m_p} (\omega^* - \omega_g) + T_f . \frac{1}{m_p} . s . \omega^* \tag{17}$$

□

This equation is of the similar form of the virtual synchronous generator described in Equation (12). The exact approximation is as follows:

$$K_I = T_f . \frac{1}{m_p} \tag{18}$$

$$K_D = \frac{1}{m_p} \tag{19}$$

Hence, the filters used for power measurements in these controllers constitute a delay which is mathematically equivalent to virtual inertia, while the droop gain is equivalent to damping. However, the traditional droop-based systems described by Equations (14) and (15) are known to have slow transient response. Moreover, the inductive grid assumption may not always be valid. Methods to improve the droop controllers, such as using virtual output impedance [16] or improving dynamic behavior of the droop scheme [14], have been proposed. In [10,66], a technique to emulate virtual inertia by a modified droop approach was also presented.

3.6. Other Topologies

Some other topologies that have been proposed in the literature are—virtual synchronous machine, referred to as "VISMA" in the literature, Institute of Electrical Power Engineering (IEPE's) topology, Kawasaki Heavy Industries (KHI) lab's topology, synchronous power controllers (SPC), virtual oscillators, inducverters, etc. The basic concept of inertia emulation remains the same in all these techniques. The VISMA topology as proposed in [67] uses *d-q* (synchronous reference frame) based mathematical model of a SG. This model when implemented in the digital controller of a power inverter replicates the dynamics of a SG. Instantaneous measurements of the grid voltage are used to compute the stator currents of the virtual machine and these currents are injected through a hysteresis current control approach using a power inverter. However, concerns with numerical instability have been reported with the VISMA model [68]. To improve robustness, a three-phase model has been proposed in [69] over a d-q based model. This is especially effective under unsymmetrical load conditions or rapid disturbances in the grid. A comparison between the VISMA algorithm implemented as a current source versus a voltage source has also been performed in [70]. The VISMA model implemented as a voltage source is referred to as IEPE's topology in the literature [28]. Instead of using voltage as input as with the VISMA topology, IEPE's topology uses the DG output current as input and generates reference voltages for the virtual machine. The IEPE topology is better suited for islanded operation, but transient currents particularly during the synchronization processes when operated in grid-connected mode can be difficult to deal with. In the KHI topology, instead of using detailed dynamic model of SG, an equivalent governor and automatic voltage regulator (AVR) model is implemented in a digital controller to generate voltage amplitude and phase reference for the virtual machine [71]. The reference is then used to generate current references based on algebraic-phasor representation of the SGs.

Another popular topology for virtual inertia implementation is the SPC as proposed in [72–74]. The general structure of the control algorithm is similar to the structure proposed in the Ise lab's topology, but instead of operating the converter as a voltage controlled system or a current controlled system, it implements a cascaded control system, with an outer voltage loop and an inner current control loop through the use of a virtual admittance. In general, such a cascaded control structure provides inherent over-current protection during severe transient operating conditions. This is lacking in other open-loop approaches such as synchronverters or the Ise lab's topology [75] described previously. SPC also avoids the discontinuities encountered in solving the mathematical models, thus making the system more robust against numerical instabilities. The nested loop structure however does entail complexity in tuning the control system parameters. Furthermore, at its core, instead of using the swing equation for inertia emulation, a second order model with an over-damped response is proposed. This helps to reduce the oscillations in the system [55]. Improved forms of this second-order model was presented in [55,76].

Inducverters [58] are one of the recent topologies that has been proposed which tries to mimic the behavior of induction generators instead of SGs. This method has the advantage of auto-synchronization without a PLL [77]. A virtual-inertia based static synchronous compensator (STATCOM) controller was proposed in [65] which behaves as synchronous condenser. The virtual inertia controller was used to exploit the fact that no PLL is required, hence providing improved voltage regulation compared to traditional STATCOMs with PLL units. Virtual oscillator controller (VOC) is another approach where, instead of mimicking synchronous/induction generators, a non-linear oscillator is implemented within the controller to synchronize DG units without any form of communication [78,79]. This approach is particularly beneficial for a grid largely dominated with DGs, as the controller is intrinsically able to maintain synchronism and share the total system load [80].

3.7. Summary of Topologies

A summary table which highlights the key features and weakness of various virtual inertia control topologies is presented in Table 1.

Table 1. Summary of Virtual Inertia Control Topologies.

Control Technique	Key Features	Weaknesses
Synchronous generator (SG) model based	• Accurate replication of SG dynamics • Frequency derivative not required • Phase locked loop (PLL) used only for synchronization	• Numerical instability concerns • Typically voltage-source implementation; no over-current protection
Swing equation based	• Simpler model compared to SG based model • Frequency derivative not required • PLL used only for synchronization	• Power and frequency oscillations • Typically voltage-source implementation; no over-current protection
Frequency-power response based	• Straightforward implementation • Typically current-source implementation; inherent over-current protection	• Instability due to PLL, particularly in weak grids • Frequency derivative required, system susceptible to noise
Droop-based approach	• Communication-less • Concepts similar to traditional droop control in SGs	• Slow transient response • Improper transient active power sharing

4. Design Procedures and Simulation Results

In this section, three of the major virtual inertia topologies were restudied in a diesel generator based remote microgrid system. The design procedures and simulation results presented are aimed to supplement the concepts of virtual inertia topologies reviewed in Section 3. Three of the topologies—the synchronverter, the Ise lab's topology, and the VSG—were implemented and their performance was studied in a common benchmark. Moreover, a procedure is provided to choose appropriate parameters for the virtual inertia systems. The three virtual inertia systems were designed in a common framework so that the different parameters used are more relatable to each other. To this end, constants in each topology were selected such that the virtual inertia system injects/absorbs the same amount of active power for a given frequency change. Furthermore, the inertial constant and the damping constant have the same proportion and were related through a time constant T_f of 0.01 s in all the simulations. This led to an inertia constant H of 1 s in all simulation cases for the virtual inertia unit. The schematic used for the virtual inertia simulation benchmark is shown in Figure 11. The generator was rated at 13 kVA, while the PV unit was rated at 6 kWp [22]. A separate, dedicated inverter unit rated at 10 kW was used as the virtual inertia unit. In all the cases, the steady-state power output from the inverter was set to 1000 W. It was assumed that, the DC side of the inverter was connected to a 400 V DC source which remained constant in all the simulations. Step changes in the load were used to emulate the change in load or PV generation in all the systems. For simplicity, the inverter was modeled as either a controlled current source or a controlled voltage source (depending on the virtual inertia topology used) neglecting the switching behavior.

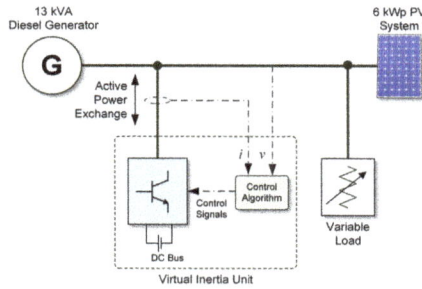

Figure 11. Schematic diagram of the virtual inertia simulation benchmark.

4.1. Design of Synchronverter Topology

The main parameters to be computed to implement a synchronverter are the moment of inertia J and the damping factor D_p. The parameter D_p can be calculated using Equation (20) from [29].

$$D_p = -\frac{\Delta T}{\Delta \omega} = -\frac{\Delta P}{\omega_g \Delta \omega} \tag{20}$$

Once D_p was calculated, the moment inertia J was computed using the desired time constant for the system, τ_f:

$$\tau_f = \frac{J}{D_p} \tag{21}$$

In this case, D_p was calculated to be 14.072 assuming ΔP of 100% (10 kW) for 0.5% change in the angular frequency (1.885 rad/s). Then for a time-constant of 0.01 s, the J value was calculated to be 0.140. The inertia constant from the synchronverter is:

$$H = \frac{J \omega_g^2}{2 P_{rated}} = 1\,s \tag{22}$$

The frequency and ROCOF of the system after a step-increase of 2 kW on the load, with and without the synchronverter, are presented in Figure 12a,b, respectively. The dip in frequency and the ROCOF of the system was reduced with the addition of the synchronverter, as expected. The additional inertia from the synchronverter increased the settling time for the frequency compared to when there was no synchronverter in the system. As shown in Figure 12c, the synchronverter increases its active power output in response to the frequency event much like the behavior of a SG.

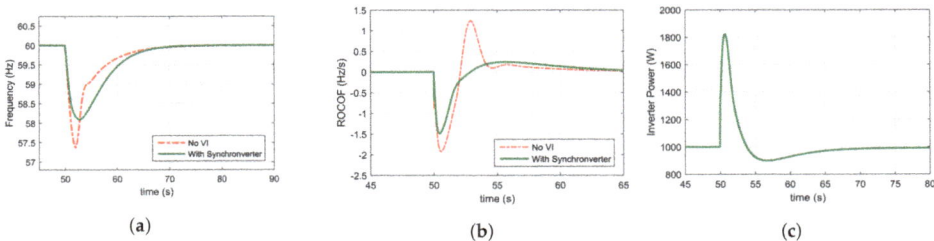

(a)   (b)   (c)

Figure 12. Simulation results from a synchronverter: (**a**) system frequency after a step-increase of 2 kW load; (**b**) ROCOF after a step-increase of 2 kW load; (**c**) increase in inverter power in response to system frequency decrease [29].

4.2. Design of Ise Lab's Topology

For the design of the Ise lab's topology, the same values for the constants J and D_p that were calculated for synchronverter in Section 4.1 were used. For the implementation of the governor model, a K value of 0.01 with a time delay T_d of 0.16 s was used. The frequency and ROCOF of the system after a step-increase of 2 kW on the load, with and without the Ise lab's system, is presented in Figure 13a,b, respectively. The dip in the frequency and the ROCOF of the system was reduced with addition of the virtual inertia unit, as expected. The additional inertia from virtual inertia system increased frequency settling time compared to the case without the virtual inertia system. The settling time, however was higher than with the synchronverter. Figure 13c shows the power injected by the inverter during the step-load increase. There is a short transient at 50 s, which was a consequence of numerical oscillation in solving the swing equation. The peak-power injected was similar to that of the synchronverter, but the time taken for the power to return to the steady-state value of 1000 W was much longer, leading to a larger energy usage from the DC side.

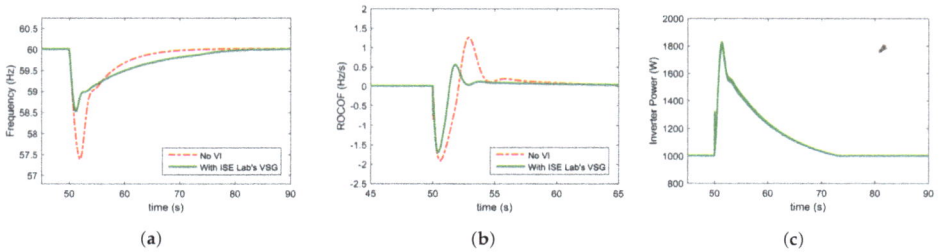

Figure 13. Simulation results from ISE lab's topology: (**a**) system frequency after a step-increase of 2 kW load; (**b**) ROCOF after a step-increase of 2 kW load; (**c**) increase in inverter power as a response to system frequency decrease [40].

4.3. Design of Virtual Synchronous Generator Topology

For implementing the VSG topology, the main parameters to be designed are the inertia constant K_I and the damping constant K_D. The parameter K_D can be calculated using:

$$K_D = \frac{\Delta P}{\omega_g \Delta \omega} \tag{23}$$

Once K_D was calculated, the inertia constant K_I was computed using the desired time constant for the system, τ_f:

$$\tau_f = \frac{K_I}{K_D} \tag{24}$$

In this case, the damping constant, K_D, was calculated to be 14.07, assuming ΔP of 100% (10 kW) for 0.5% change in the angular frequency (1.885 rad/s). Then, for a time-constant of 0.01 s, the K_I value was calculated to be 0.14. The inertia constant from the VSG is:

$$H = \frac{K_I \omega_g^2}{2 P_{rated}} = 1s \tag{25}$$

The frequency and ROCOF of the system after a step-increase of 2 kW on the load, with and without the VSG, is presented in Figure 14a,b, respectively. The dip in frequency and the ROCOF of the system was reduced with addition of the VSG, as expected. As with the previous cases, the additional inertia from the VSG slowed the system down, and the settling time for the frequency was increased compared to the case without virtual inertia. The peak-power injected was slightly higher than that of the synchronverter and Ise lab's topology. However, the time taken for the power to return to the steady value of 1000 W was much longer than for the synchronverter leading to a larger energy usage from the DC side.

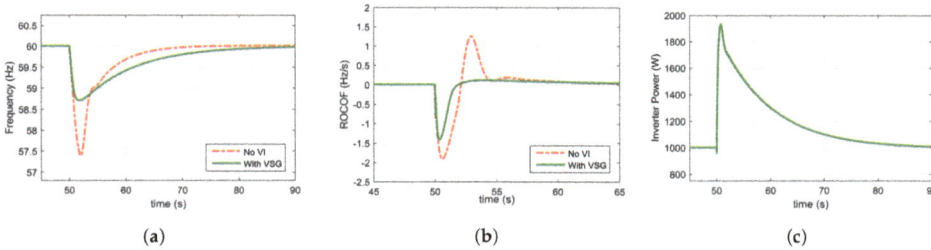

Figure 14. Simulation results from a virtual synchronous generator: (**a**) system frequency after a step-increase of 2 kW load; (**b**) ROCOF after a step-increase of 2 kW load; (**c**) increase in inverter power as a response to system frequency decrease [45,46].

4.4. Summary of Simulations

The simulation results are summarized in Table 2 in terms of parameters like the minimum frequency, maximum ROCOF, settling time, peak power, and energy exchange. The settling time is defined here as the time required for the frequency to return to and stay within ±0.25 Hz of the final steady-state frequency after a disturbance. The energy exchange was calculated over the time period where the inverter exchanges power with the system. With all three topologies, the minimum frequency and ROCOF were reduced by similar amounts. The peak power delivered by the inverter varied slightly, with the highest value of 1929 W for the VSG topology. The most pronounced differences were in the settling time for the frequency and the energy exchange. Compared to systems with no virtual inertia, the settling time has increased in all three cases. This was expected as adding virtual inertia slows down the frequency dynamics. The ISO8528-5 standard for generators sets recommends a settling time of 10 s [20]. The settling time, however, increased to 13.2 s with synchronverter and an even higher value of 17.7 s and 17.9 s with the Ise lab's and VSG respectively. This led to a relatively higher energy exchange in these two topologies of 3.8 Wh and 4.9 Wh compared to that 0.8 Wh with the synchronverter. Moreover, there was a short-energy recovery period in the power plot of the synchronverter as seen in Figure 12c which led to a lower energy exchange estimate for the synchronverter.

Table 2. Performance comparison of systems without virtual inertia (VI), and VI implemented through synchronverter, Ise lab's and virtual synchronous generator (VSG) topologies.

Parameter	No VI	Synchronverter	Ise Lab	VSG
Minimum Frequency	57.3 Hz	58.1 Hz	58.6 Hz	58.3 Hz
Maximum ROCOF	1.9 Hz/s	1.5 Hz/s	1.6 Hz/s	1.7 Hz/s
Settling time	11.3 s	13.2 s	17.7 s	17.9 s
Peak power delivered	0 W	1825 W	1800 W	1929 W
Energy exchanged	0 Wh	0.8 Wh	3.8 Wh	4.9 Wh

Therefore, by appropriate selection of the parameters for the topologies through the time constant T_f and/or the inertia constant H, similar inertial response can be achieved in terms of frequency deviation reduction and power exchange from the inverter. Based on the topology, the exact dynamics represented by the system may vary. The selection of a particular topology depends on the application and the desired level of replication of the dynamics of the SG. Topologies like the synchronverter and the Ise lab's topology may be more suitable for isolated power system as they can operate autonomously as grid forming units, as well as for reasons discussed in Section 3. The VSG topology on the other hand behaves more like a grid following unit with added inertial response capabilities and is more suited towards interconnected operations. The synchronverter or Ise lab's topology are more suitable for a closer approximation of SG dynamics. If the main aim, however, is to make the DG unit responsive to frequency changes, the VSG approach provides a far simpler implementation.

5. Second Generation: Optimization of Virtual Inertia Systems

The first generation of virtual inertia systems in the literature focused on developing novel topologies for emulation of inertia using power electronic converters. These topologies have matured since as pointed out in Section 3. Recently, the field is more focused towards improving and optimizing the performance of these topologies from the point-of-view of enhanced dynamics, stability, and minimizing energy storage requirements.

5.1. Second Generation of Synchronverters

Improved versions of the synchronverter have been proposed in [81,82] which makes the synchronverter more robust and allows for an more accurate dynamic representation of SGs. One of the main improvements (among others) in [82] is virtually increasing the filter inductance of the synchronverter, which improved the stability compared to the original synchronverter. This modification allowed for an improved control over the response speed of the frequency loop proposed in [29]. In a similar theme, an auxiliary loop around the frequency-loop was proposed in [83] which allowed for a free control of the response speed of synchronverter. This auxiliary loop did not affect the steady-state drooping mechanism of the synchronverter which is very desirable. By changing the inertia constant J and a different tunable constant D_f, the desired response speed was achieved. In [84], a synchronverter with analytically determined bounds for frequency and voltage was introduced. In traditional synchronverters, saturation units were employed for this purpose, but such an approach can lead to instability due to wind-up. Instead, analytically determined bounds based on the system parameters were proposed to improve stability.

5.2. Second Generation of Ise Lab's Topology

In the traditional Ise lab's topology, active power oscillation during the inertia emulation has been identified as one of the major concerns [41]. Typically, during a frequency event, the DG unit needs to release/absorb a high amount of power, which may exceed their power ratings. This is not a problem for conventional SGs as they have inherently overrated operation capabilities. However, in the case of inverters, the switches have to be over-sized to handle such peak power, leading to an increase in inverter size and, consequently, cost [36]. In [41], an alternating moment of inertia emulation approach was proposed to make the system less susceptible to such oscillations. The J parameter was changed based on the relative "virtual angular velocity" and its rate of change. The proposed alternating moment of inertia approach not only stabilized the system under consideration, but other nearby virtual inertia units as well. Similarly, in [85] another technique of adjusting the "virtual stator reactance" of the virtual inertia unit has also been proposed to reduce such active power oscillations. This approach was somewhat similar to the approach described for synchronverters in [82]. The technique was also found to aid in proper transient active power sharing when operating multiple virtual inertia units in a microgrid environment. In [86], a particle swarm optimization technique was developed to properly tune the parameters of the system and achieve smooth transitions after a disturbance when operating multiple virtual inertia units.

5.3. Second Generation of Virtual Synchronous Generators

In terms of improvement in VSG topologies, some researchers have developed techniques to try to minimize the frequency nadirs/peaks in the system at the expense of higher energy usage and peak transient power exchange through the virtual inertia systems [87,88]. Other researchers, meanwhile, have focused on reducing the energy storage requirements and limiting peak transient power in virtual inertia systems even though it leads to slightly higher frequency nadirs/peaks [89,90]. A self-tuning

VSG was developed in [89] using an online optimization technique to tune the K_I and K_D parameters of the VSG control algorithm (described in Section 3.4) to minimize the frequency excursions, the ROCOF, and the power flow through the ESS. Although the frequency excursions were slightly higher in the case of the self-tuning algorithm, the power flow through the ESS was reduced by 58%. Moreover, the technique used less energy per Hz of frequency reduction than a constant parameter VSG.

On a similar note of energy saving, an online neural-network based controller was proposed in [90,91]. It used an adaptive dynamic programming (ADP) based approach to optimize the system and minimize energy usage while limiting the transient power. The controller supplemented the power references generated by the main VSG algorithm P_{VSG} with a supplementary signal P_{ADP} to give the total reference $P_{VSG,TOTAL}$ as shown in Figure 15a. The aim of this supplementary signal was to improve the dynamics of virtual inertia. The proposed ADP controller used a neural network structure with two different networks—an action network and a critic network as shown in Figure 15b. The idea behind the design of the critic network was to adapt its weight such that the optimal cost function $J^*(X(t))$ satisfies the Bellman principle of optimally as given by:

$$J^*(X(t)) = \substack{min \\ u(t)} \left\{ J^*(X(t+1)) + r(X(t)) - U_c \right\} \tag{26}$$

where, $r(t)$ is the reinforcement signal for the critic network and U_c is a heuristic term used to balance. The input to the supplementary ADP controller was the state vector $X(t)$ where the elements were the frequency error and the one and two time-step delayed frequency error signals. Based on a reinforcement learning approach, the ADP controller generated auxiliary power reference signals P_{ADP} to return the frequency back to its steady-state value faster and as a consequence reduced the energy exchange as explained in [90,91]. The main concern with adding virtual inertia to the system is that it can increase the frequency settling time, leading to increased energy exchange from the ESS, which subsequently shortens the life of the ESS. The online controller was able to reduce the frequency settling time and the transient peak power. Figure 16a shows the frequency of a PV-hydro system under step load changes with and without the ADP controller. The frequency excursion was slightly higher than using constant parameter VSG, but there was a reduction in the frequency settling time. This led to lower energy usage and lower transient power as observable in Figure 16b. Table 3 summarizes the improvement achieved through the ADP-based virtual inertia controller.

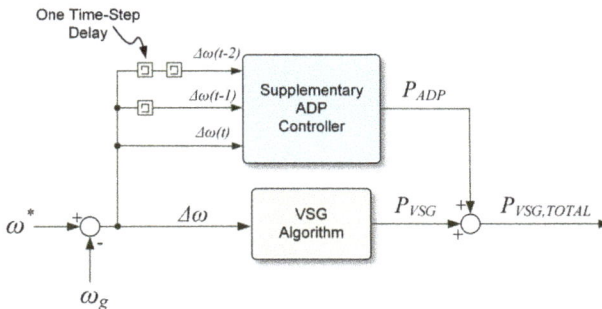

(a)

Figure 15. *Cont.*

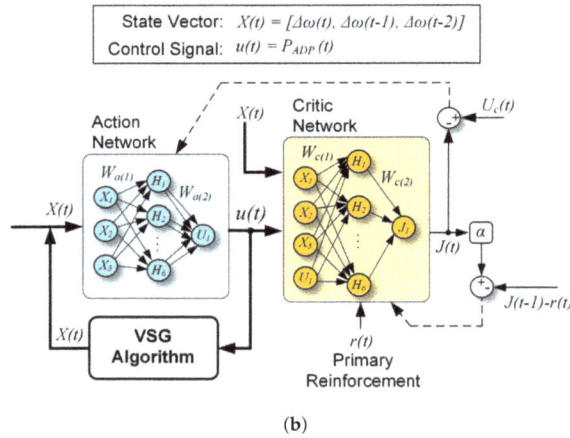

Figure 15. Modified virtual synchronous generator (VSG) using adaptive dynamic programming (ADP) (a) overall schematic of the controller; (b) the action and critic neural network based structure.

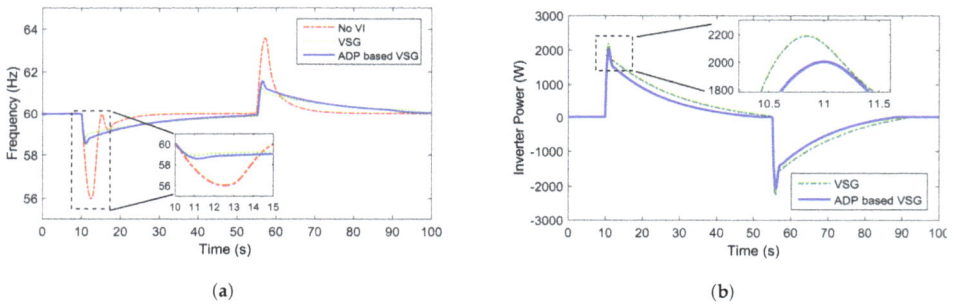

Figure 16. Comparison of traditional virtual synchronous generator (VSG) controller with the online learning based controller: (a) frequency of the system for step load changes; (b) power exchange with the system (Adapted from [91]).

Table 3. Performance comparison of the system without virtual inertia (VI), simple virtual synchronous generator (VSG) based and adaptive dynamic programming (ADP) based VSG (Data from [91]).

Parameter	No VI	Simple VSG	ADP Based VSG
Peak Power for Event A	0 W	2184 W	1979 W
Settling time for Event A	12.6 s	35.1 s	31.3 s
Peak Power for Event B	0 W	−2235 W	−2029 W
Settling time for Event B	11.1 s	29.1 s	26.6 s
Energy delivered (Wh)	0 Wh	8.2 Wh	6.2 Wh
Net energy exchanged (Wh)	0 Wh	1.6 Wh	0.9 Wh

A similar online learning controller was proposed for virtual inertia implementation in a double fed induction generator (DFIG) based system in [87]. In this case, the controller was trained so as to restrict the frequency excursions to a minimum while maintaining the rotor speed of the DFIG within a safe operating range, rather than saving the energy flow from ESS. Other techniques to optimize the virtual inertia have been proposed in [88] using Linear-quadratic-regulator (LQR) and in [92] using fuzzy logic to minimize frequency deviations and ROCOF.

162

6. Challenges and Future Research Directions

6.1. Virtual Inertia as an Ancillary Service

Many research works have proposed the possibility of using virtual inertia as an ancillary service to improve frequency stability of large power grids. In [93], a control scheme to integrate DC microgrids as virtual inertia emulating units in the traditional AC grid has been presented. With the control scheme, the resources within the DC microgrid can be dispatched as an ancillary service for inertial response. Another major source of under-utilized energy lies in modern data centers. Data centers need a high degree of reliability, and as a result large amounts of backup energy storage which are unused during normal operating conditions. Research work in [94,95] have shown methods to utilize these resources using demand response techniques. This concept can be extended to use data center resources for virtual inertia implementation. Virtual inertia based interfaces, as mentioned in [93], can be integrated with data center resources for frequency regulation. A unit commitment model that combines system inertia from the conventional plants, and the virtual inertia from wind plants into system scheduling has been presented in [96], which allows for an economic analysis of the virtual inertia system.

Modern wind farms are already obligated through various laws and regulations to provide inertial ancillary services [97–100]. The uncaptured inertia in wind turbines, referred to as "hidden inertia", can be captured through the techniques described in previous sections. Commercial wind turbine manufacturers, like WindINERTIA [101] and ENERCON [102], already provide virtual inertia response. Moreover, leading inverter manufacturers like FREQCON, Schneider Electric, and ABB already provide out-of-the-box inertial response capabilities. Using electric vehicles (EVs) to provide ancillary services has become a popular research topic [103]. Typically the control algorithm of the bidirectional converters in EVs can be modified for virtual inertia implementation [104,105].

6.2. Inertia Estimation

Research has been conducted in [106] to estimate the total inertia constant of the power system. The research was aimed at determining spinning reserve requirements for the power system. However, virtual inertia emulated using ESSs and RESs is not going to be constant as in the case of traditional synchronous generation. The available inertia in the system will depend upon whether RES units are online or not, and resource availability (wind speed, irradiance, and state of charge in case of ESS) [107]. System inertia estimation is thus going to be critical for planning purposes for system operators in the future power system with high RES penetrations. Furthermore, such estimates can provide helpful insights into the stable real-time operation of a power system. Inertia estimation using frequency transients measured using synchronized phasor measurement units (PMUs) was proposed in [107,108]. In [109], a method to estimate the inertial response of power system under high wind penetration based on the swing equation is presented. Accurate detection of frequency events and precise ROCOF measurements are critical for proper inertia estimation [110]. In the context of modern power systems with RES units participating in the inertial response, the inertia of the system will also largely depend on the RES resource availability at any given time as well. So, PV and wind forecasts data can be used to complement and further improve the inertia estimation techniques described before. Accurate inertia estimation methods will help setup a framework for system operator's to procure inertial services.

6.3. Improved Modeling, Control and Aggregation of Virtual Inertia Systems

Most research has focused on specific implementations of virtual inertia and the broader impacts of inertial response. Current literature, lacks accurate, mathematical models which represent the dynamics of the system. Such models are essential for parameter tuning and understanding the operational behavior when virtual inertia systems are interconnected to the power system. In [75,111], a small-signal model for a virtual inertia system has been developed. The model was used to identify critical operating modes through Eigenvalue analysis, and a technique to assess the sensitivity of the system to the parameter gains has been demonstrated. Similarly, a small-signal model of a synchronverter was developed in [32]. Such a model aids in improved tuning of the controller gains and provides granular control over how the overall system needs to be operated. An analytical approach to study the effect of microgrids with high RES penetration on the frequency stability has been described in [112]. Performance indices completely independent of the test system have also been proposed to better facilitate impact analysis. The behavior and coordination between virtual inertia systems and existing SGs are critical topics for further research. In the future, with numerous virtual inertia units, the coordinated and aggregated operation, and optimal placement will also be important research questions.

6.4. Market Structure for Virtual Inertia Systems

Currently no market for virtual inertia nor for inertia from conventional SGs exists. SGs and some loads in the power system inherently provide inertial response and are treated as a free resource. As the power system becomes inverter-dominated, the inertial requirements will become a valuable tradeable commodity, and generating units will demand financial compensation. A market-based approach can be a cost effective solution to ensure sufficiency of inertial services in the future power market [113]. The inertial response can be provided by wind turbines or even PV systems with inherent storage technologies [114]. Schemes to operate PV systems below their MPP with reserve for inertial response is also a possible option with the suitable market for such resources. A scheme to trade inertia is presented in [11]. Furthermore, the paper argues inertia should not be traded in terms of power or energy, but rather in terms of an inertia metric. A unit commitment framework for fast frequency services in the power system with transient stability constraints representing the dynamic performance requirements was proposed in [115]. It was shown that additional inertia prevented expensive units being committed post-frequency event and reduced the overall system production cost in a power system. Other papers propose a penalty factor for generators that do not provide inertial response, but so far there is no clear structure on how the inertia market should operate and is an open research area.

One method that deserves further exploration is deploying inertia as "service" for power quality. For instance, as a microgrid operator, one can offer inertial services based on certain criteria such as maximum allowable ROCOFs and/or frequency deviation. The Quality of Service (QoS) metrics which have been proposed for cloud computing services (e.g., [116]) can be garnered for power systems to measure the power quality in terms on inertial response availability. The quality may be assessed in terms of response time after a frequency disturbance and/or inertia made available. This will foster a framework for microgrid operators to incorporate inertial response services in the system based on the requirements of its end-users.

6.5. Energy Storage Resources for Virtual Inertia Systems

Typically, capacitors and batteries have been proposed as ESSs for dynamic frequency control using power electronic converters [45,50,72,82]. In [117], an ultra-capacitor based ESS is proposed to reduce the impact of RESs variability of frequency stability of an isolated power system. However, these energy resources often incorporate prohibitive cost investments, and because fast-frequency needs to compensated by the virtual inertia systems it may effect the lifetime of the ESS. As a solution, a parallel combination of batteries and ultra-capacitors was proposed in [21] which significantly reduced the impact of high frequency dynamics on the batteries as the ultra-capacitors supplied the high frequency components. This also allowed for a cheaper and smaller battery unit [118]. Flywheel based energy storage for virtual inertia was proposed in [119]. Novel solar panel technologies with inherent storage capabilities could be another way of providing inertia through PV systems [114]. Recently, researchers have started to focus on alternate means of energy resource for virtual inertia. One of the main areas that is gaining attention is the so-called "thermal-inertia" of heating, ventilation, and air conditioning (HVAC) systems of commercial buildings. As discussed in [120–122], the power consumption of the power electronics based HVAC units can be controlled to provide inertial response while ensuring that the customer comfort is not effected. Similarly, the large HVAC installation in data centers could be another potential to tap for inertial response in the future grid with large scale integration of RES units.

7. Conclusions

This paper presented a literature review of virtual inertia systems in the modern power system under high RES penetration. Numerous topologies for virtual inertia implementation, which constitutes the "first generation" of virtual inertia systems, were identified. It was shown that, fundamentally, the objective of all the topologies is to provide dynamic frequency response through power electronic converters. The appropriate topology can be selected based on the required architecture (current source or voltage source implementation) and desired level of sophistication in emulating the exact behavior of SGs. For example, for replication of the exact dynamics of SGs, topologies such as the synchronverter, VISMA and inducverters can be used. More simplistic topologies like Ise lab's topology, SPC can be used if an approximate replication is sufficient. The VSG approach, on the other hand, is more suitable when the objective is to provide just the dynamic frequency response without emulating the exact behavior of SGs. An important takeaway through the literature review was that the droop based controllers, which were regarded as separate control method for inverter systems, are in fact fundamentally similar to virtual inertia systems as formalized by the literature pointed out.

Next, the second generation of virtual inertia systems with focus on optimization of existing virtual inertia topologies were reviewed. Such algorithms can prevent degradation of ESS lifetime and allow reduced curtailment of RES units that participate in inertial response. Furthermore, the enhancements help in improved dynamics and overall stability. Some of the challenges and possible areas where further research is required were also discussed. The current state-of-art of topics such as inertia estimation, improved controls and aggregation techniques, the virtual inertia market, and ESS for virtual inertia systems were also presented. This was followed by a discussion on possible research directions on these topics.

Acknowledgments: The authors would like to thank Microsoft Inc. and South Dakota Board of Regents (SDBOR) for the financial support. The authors are also thankful towards Mr. Jason Sternhagen for his help in proofreading the paper.

Author Contributions: Ujjwol Tamrakar replicated and analyzed the model of various topologies available in the literature; Ujjwol Tamrakar and Dipesh Shrestha developed one of the case studies; Ujjwol Tamrakar, Dipesh Shrestha, and Reinaldo Tonkoski analyzed the topologies and formulated the discussions in the manuscript. Ujjwol Tamrakar, Dipesh Shrestha, and Manisha Maharjan wrote the manuscript; Bishnu P. Bhattarai and Timothy M. Hansen contributed on the formulation of the discussion and provided their comprehensive feedback on the paper.

Conflicts of Interest: The authors declare no conflict of interest.

Abbreviations

The following abbreviations are used in this manuscript:

ADP	Adaptive Dynamic Programming
DFIG	Double Fed Induction Generator
DG	Distributed Generation
ENTSO-E	European Network of Transmission System Operators for Electricity
ERCOT	Electricity Reliability Council of Texas
ESS	Energy Storage System
HVAC	Heating, Ventilation and Air Conditioning
IEEE	Institute of Electrical and Electronic Engineers

IEPE	Institute of Electrical Power Engineering
LQR	Linear Quadratic Regulator
KHI	Kawasaki Heavy Industries
MPP	Maximum Power Point
NERC	North American Electric Reliability Corporation
PI	Proportional-Integral
PLL	Phase Locked Loop
PWM	Pulse Width Modulation
QoS	Quality of Service
RES	Renewable Energy System
ROCOF	Rate of Change of Frequency
SG	Synchronous Generator
SPC	Synchronous Power Controller
STATCOM	Static Synchronous Compensator
VI	Virtual Inertia
VISMA	Virtual Synchronous Machine
VOC	Virtual Oscillator Controller
VSG	Virtual Synchronous Generator
UFLS	Under Frequency Load Shedding

References

1. U.S. Energy Information Administration. *Annual Energy Outlook 2017*; U.S. Department of Energy: Washington, DC, USA, 2017. Available online: https://www.eia.gov/outlooks/aeo/pdf/0383(2017).pdf (accessed on 21 June 2017).
2. The Sunshot Initiative. Available online: http://www.webcitation.org/6pV7YpHo2 (accessed on 5 April 2017).
3. IEA PVPS. *Trends in 2016 in Photovoltaic Applications*; T1-30:2016; IEA PVPS: Paris, France, 2016. Available online: http://iea-pvps.org/fileadmin/dam/public/report/national/Trends_2016_-_mr.pdf (accessed on 21 June 2017).
4. Kroposki, B.; Johnson, B.; Zhang, Y.; Gevorgian, V.; Denholm, P.; Hodge, B.M.; Hannegan, B. Achieving a 100% Renewable Grid: Operating Electric Power Systems with Extremely High Levels of Variable Renewable Energy. *IEEE Power Energy Mag.* **2017**, *15*, 61–73.
5. Hussein, M.M.; Senjyu, T.; Orabi, M.; Wahab, M.A.; Hamada, M.M. Control of a stand-alone variable speed wind energy supply system. *Appl. Sci.* **2013**, *3*, 437–456.
6. Yan, R.; Saha, T.K.; Modi, N.; Masood, N.A.; Mosadeghy, M. The combined effects of high penetration of wind and PV on power system frequency response. *Appl. Energy* **2015**, *145*, 320–330.
7. Electricity Reliability Council of Texas (ERCOT). *Future Ancillary Services in ERCOT*; ERCOT: Taylor, TX, USA, 2013. Available online: http://www.ercot.com/content/news/presentations/2014/ERCOT_AS_Concept_Paper_Version_1.1_as_of_11-01-13_1445_black.pdf (accessed on 21 June 2017).
8. Matevosyan, J.; Sharma, S.; Huang, S.H.; Woodfin, D.; Ragsdale, K.; Moorty, S.; Wattles, P.; Li, W. Proposed future Ancillary Services in Electric Reliability Council of Texas. In Proceedings of the IEEE PowerTech, Eindhoven, The Netherlands, 29 June–2 July 2015; pp. 1–6.
9. Poolla, B.K.; Bolognani, S.; Dorfler, F. Optimal placement of virtual inertia in power grids. In Proceedings of the American Control Conference, Boston, MA, USA, 6–8 July 2016.
10. De Vyver, J.V.; Kooning, J.D.M.D.; Meersman, B.; Vandevelde, L.; Vandoorn, T.L. Droop Control as an Alternative Inertial Response Strategy for the Synthetic Inertia on Wind Turbines. *IEEE Trans. Power Syst.* **2016**, *31*, 1129–1138.
11. Thiesen, H.; Jauch, C.; Gloe, A. Design of a System Substituting Today's Inherent Inertia in the European Continental Synchronous Area. *Energies* **2016**, *9*, 582.
12. Gurung, A.; Galipeau, D.; Tonkoski, R.; Tamrakar, I. Feasibility study of Photovoltaic-hydropower microgrids. In Proceedings of the 5th International Conference on Power and Energy Systems (ICPS), Kathmandu, Nepal, 28–30 October 2014; pp. 1–6.

13. Tamrakar, U.; Galipeau, D.; Tonkoski, R.; Tamrakar, I. Improving transient stability of photovoltaic-hydro microgrids using virtual synchronous machines. In Proceedings of the IEEE Eindhoven PowerTech, Eindhoven, The Netherlands, 29 June–2 July 2015; pp. 1–6.

14. Guerrero, J.M.; de Vicuna, L.G.; Matas, J.; Castilla, M.; Miret, J. A wireless controller to enhance dynamic performance of parallel inverters in distributed generation systems. *IEEE Trans. Power Electron.* **2004**, *19*, 1205–1213.

15. Hatziargyriou, N.; Asano, H.; Iravani, R.; Marnay, C. Microgrids. *IEEE Power Energy Mag.* **2007**, *5*, 78–94.

16. Kim, J.; Guerrero, J.M.; Rodriguez, P.; Teodorescu, R.; Nam, K. Mode Adaptive Droop Control With Virtual Output Impedances for an Inverter-Based Flexible AC Microgrid. *IEEE Trans. Power Electron.* **2011**, *26*, 689–701.

17. Under Frequency Load Shedding. Available online: http://www.nerc.com/pa/RAPA/ri/Pages/UnderFrequencyLoadShedding.aspx (accessed on 25 May 2017).

18. North American Electric Reliability Corporation (NERC). *Frequency Response Initiative Report: The Reliability Role of Frequency Response*; NERC: Atlanta, GA, USA, 2012. Available online: http://www.nerc.com/docs/pc/FRI_Report_10-30-12_Master_w-appendices.pdf (accessed on 21 June 2017).

19. Voltage Characteristics of Electricity Supplied by Public Distribution Systems. Available online: http://www2.schneider-electric.com/library/SCHNEIDER_ELECTRIC/SE_LOCAL/APS/204836_1312/DraftStandard0026rev2-DraftEN501602005-05.pdf (accessed on 21 June 2017).

20. ISO 8528-5:2005 Standard: Reciprocating Internal Combustion Engine Driven Alternating Current Generating Sets—Part 5: Generating Sets. 2005. Available online: https://www.iso.org/standard/39047.html (accessed on 21 June 2017).

21. Torres, M.; Lopes, L.A. Virtual synchronous generator: A control strategy to improve dynamic frequency control in autonomous power systems. *Energy Power Eng.* **2013**, *5*, 2A:1–2A:7. Available online: http://file.scirp.org/Html/5-6201497_30602.htm (accessed on 29 April 2017).

22. Shrestha, D.; Tamrakar, U.; Ni, Z.; Tonkoski, R. Experimental Verification of Virtual Inertia in Diesel Generator based Microgrids. In Proceedings of the 18th Annual International Conference on Industrial Technology (ICIT), Toronto, ON, Canada, 22–25 March 2017; pp. 95–100.

23. Rahmann, C.; Castillo, A. Fast Frequency Response Capability of Photovoltaic Power Plants: The Necessity of New Grid Requirements and Definitions. *Energies* **2014**, *7*, 6306–6322.

24. Chang-Chien, L.R.; Lin, W.T.; Yin, Y.C. Enhancing Frequency Response Control by DFIGs in the High Wind Penetrated Power Systems. *IEEE Trans. Power Syst.* **2011**, *26*, 710–718.

25. Tielens, P.; Hertem, D.V. The relevance of inertia in power systems. *Renew. Sustain. Energy Rev.* **2016**, *55*, 999–1009.

26. Beck, H.P.; Hesse, R. Virtual synchronous machine. In Proceedings of the 9th International Conference on Electrical Power Quality and Utilisation, Barcelona, Spain, 9–11 October 2007; pp. 1–6.

27. Gonzalez-Longatt, F.; Chikuni, E.; Rashayi, E. Effects of the Synthetic Inertia from wind power on the total system inertia after a frequency disturbance. In Proceedings of the IEEE International Conference on Industrial Technology (ICIT), Cape Town, South Africa, 25–28 February 2013; pp. 826–832.

28. Bevrani, H.; Ise, T.; Miura, Y. Virtual synchronous generators: A survey and new perspectives. *Int. J. Electr. Power Energy Syst.* **2014**, *54*, 244–254.

29. Zhong, Q.C.; Weiss, G. Synchronverters: Inverters That Mimic Synchronous Generators. *IEEE Trans. Ind. Electron.* **2011**, *58*, 1259–1267.

30. Zhong, Q.C. Virtual Synchronous Machines: A unified interface for grid integration. *IEEE Power Electron. Mag.* **2016**, *3*, 18–27.

31. Kundur, P.; Balu, N.J.; Lauby, M.G. *Power System Stability and Control*; McGraw-Hill: New York, NY, USA, 1994.

32. Piya, P.; Karimi-Ghartemani, M. A stability analysis and efficiency improvement of synchronverter. In Proceedings of the IEEE Applied Power Electronics Conference and Exposition (APEC), Long Beach, CA, USA, 20–24 March 2016; pp. 3165–3171.

33. Zhong, Q.C.; Hornik, T. Sinusoid-Locked Loops. In *Control of Power Inverters in Renewable Energy and Smart Grid Integration*; John Wiley & Sons, Ltd.: Hoboken, NJ, USA, 2012; pp. 379–392.

34. Ferreira, R.V.; Silva, S.M.; Brandao, D.I.; Antunes, H.M.A. Single-phase synchronverter for residential PV power systems. In Proceedings of the 17th International Conference on Harmonics and Quality of Power (ICHQP), Belo Horizonte, Brazil, 16–19 October 2016; pp. 861–866.

35. Shinnaka, S. A Robust Single-Phase PLL System With Stable and Fast Tracking. *IEEE Trans. Ind. Appl.* **2008**, *44*, 624–633.

36. Zhang, L.; Harnefors, L.; Nee, H.P. Power-Synchronization Control of Grid-Connected Voltage-Source Converters. *IEEE Trans. Power Syst.* **2010**, *25*, 809–820.

37. Wang, S.; Hu, J.; Yuan, X. Virtual Synchronous Control for Grid-Connected DFIG-Based Wind Turbines. *IEEE J. Emerg. Sel. Top. Power Electron.* **2015**, *3*, 932–944.

38. Zhong, Q.C.; Nguyen, P.L.; Ma, Z.; Sheng, W. Self-Synchronized Synchronverters: Inverters Without a Dedicated Synchronization Unit. *IEEE Trans. Power Electron.* **2014**, *29*, 617–630.

39. Ma, Z.; Zhong, Q.C.; Yan, J.D. Synchronverter-based control strategies for three-phase PWM rectifiers. In Proceedings of the 7th IEEE Conference on Industrial Electronics and Applications (ICIEA), Singapore, 18–20 July 2012; pp. 225–230.

40. Sakimoto, K.; Miura, Y.; Ise, T. Stabilization of a power system with a distributed generator by a Virtual Synchronous Generator function. In Proceedings of the 8th International Conference on Power Electronics (ECCE Asia), Jeju, Korea, 30 May–3 June 2011; pp. 1498–1505.

41. Alipoor, J.; Miura, Y.; Ise, T. Power System Stabilization Using Virtual Synchronous Generator With Alternating Moment of Inertia. *IEEE J. Emerg. Sel. Top. Power Electron.* **2015**, *3*, 451–458.

42. Liu, J.; Miura, Y.; Ise, T. Dynamic characteristics and stability comparisons between virtual synchronous generator and droop control in inverter-based distributed generators. In Proceedings of the International Power Electronics Conference (IPEC-Hiroshima 2014-ECCE ASIA), Hiroshima, Japan, 18–21 May 2014; pp. 1536–1543.

43. Sakimoto, K.; Miura, Y.; Ise, T. Characteristics of Parallel Operation of Inverter-Type Distributed Generators Operated by a Virtual Synchronous Generator. *Electr. Eng. Jpn.* **2015**, *192*, 9–19.

44. Torres, M.; Lopes, L.A.C. Virtual synchronous generator control in autonomous wind-diesel power systems. In Proceedings of the IEEE Electrical Power & Energy Conference (EPEC), Montreal, QC, Canada, 22–23 October 2009; pp. 1–6.

45. Van Wesenbeeck, M.P.N.; de Haan, S.W.H.; Varela, P.; Visscher, K. Grid tied converter with virtual kinetic storage. In Proceedings of the IEEE Bucharest PowerTech, Bucharest, Romania, 28 June–2 July 2009; pp. 1–7.

46. Van, T.V.; Visscher, K.; Diaz, J.; Karapanos, V.; Woyte, A.; Albu, M.; Bozelie, J.; Loix, T.; Federenciuc, D. Virtual synchronous generator: An element of future grids. In Proceedings of the IEEE Innovative Smart Grid Technologies Conference Europe (ISGT Europe), Gothenberg, Sweden, 11–13 October 2010; pp. 1–7.

47. Wen, B.; Boroyevich, D.; Burgos, R.; Mattavelli, P.; Shen, Z. Small-Signal Stability Analysis of Three-Phase AC Systems in the Presence of Constant Power Loads Based on Measured d-q Frame Impedances. *IEEE Trans. Power Electron.* **2015**, *30*, 5952–5963.

48. Tamrakar, U.; Tonkoski, R.; Ni, Z.; Hansen, T.M.; Tamrakar, I. Current control techniques for applications in virtual synchronous machines. In Proceedings of the 6th IEEE International Conference on Power Systems (ICPS), New Delhi, India, 4–6 March 2016; pp. 1–6.

49. Malesani, L.; Tomasin, P. PWM current control techniques of voltage source converters—A survey. In Proceedings of the International Conference on Industrial Electronics, Control, and Instrumentation (IECON '93), Maui, HI, USA, 15–19 November 1993; pp. 670–675.

50. Driesen, J.; Visscher, K. Virtual synchronous generators. In Proceedings of the 9th IEEE Power & Energy Society General Meeting, Pittsburgh, PA, USA, 20–24 July 2008; pp. 1–6.

51. Karapanos, V.; de Haan, S.; Zwetsloot, K. Real time simulation of a power system with VSG hardware in the loop. In Proceedings of the 37th Annual Conference of the IEEE Industrial Electronics Society (IECON), Melbourne, Australia, 7–10 November 2011; pp. 3748–3754.

52. Thong, V.V.; Woyte, A.; Albu, M.; Hest, M.V.; Bozelie, J.; Diaz, J.; Loix, T.; Stanculescu, D.; Visscher, K. Virtual synchronous generator: Laboratory scale results and field demonstration. In Proceedings of the IEEE Bucharest PowerTech, Bucharest, Romania, 28 June–2 July 2009; pp. 1–6.

53. Morren, J.; Pierik, J.; de Haan, S.W. Inertial response of variable speed wind turbines. *Electr. Power Syst. Res.* **2006**, *76*, 980–987.

54. Arani, M.F.M.; El-Saadany, E.F. Implementing Virtual Inertia in DFIG-Based Wind Power Generation. *IEEE Trans. Power Syst.* **2013**, *28*, 1373–1384.

55. Zhang, W.; Cantarellas, A.M.; Rocabert, J.; Luna, A.; Rodriguez, P. Synchronous Power Controller With Flexible Droop Characteristics for Renewable Power Generation Systems. *IEEE Trans. Sustain. Energy* **2016**, *7*, 1572–1582.

56. Blaabjerg, F.; Teodorescu, R.; Liserre, M.; Timbus, A.V. Overview of Control and Grid Synchronization for Distributed Power Generation Systems. *IEEE Trans. Ind. Electron.* **2006**, *53*, 1398–1409.

57. European Network of Transmission System Operators for Electricity (ENTSO-E). *Need for Synthetic Inertia (SI) for Frequency Regulation*; ENTSO-E: Brussels, Belgium, 2017. Available online: https://consultations.entsoe.eu/system-development/entso-e-connection-codes-implementation-guidance-d-3/user_uploads/igd-need-for-synthetic-inertia.pdf (accessed on 21 June 2017).

58. Ashabani, M.; Freijedo, F.D.; Golestan, S.; Guerrero, J.M. Inducverters: PLL-Less Converters With Auto-Synchronization and Emulated Inertia Capability. *IEEE Trans. Smart Grid* **2016**, *7*, 1660–1674.

59. D'Arco, S.; Suul, J.A. Virtual synchronous machines- Classification of implementations and analysis of equivalence to droop controllers for microgrids. In Proceedings of the IEEE Grenoble Conference, Grenoble, France, 16–20 June 2013; pp. 1–7.

60. Midtsund, T.; Suul, J.A.; Undeland, T. Evaluation of current controller performance and stability for voltage source converters connected to a weak grid. In Proceedings of the IEEE 2nd International Symposium on Power Electronics for Distributed Generation Systems, Hefei, China, 16–18 June 2010; pp. 382–388.

61. Svensson, J. Synchronisation methods for grid-connected voltage source converters. *IEE Proc. Gener. Transm. Distrib.* **2001**, *148*, 229–235.

62. Katiraei, F.; Iravani, M.R. Power Management Strategies for a Microgrid With Multiple Distributed Generation Units. *IEEE Trans. Power Syst.* **2006**, *21*, 1821–1831.

63. Pogaku, N.; Prodanovic, M.; Green, T.C. Modeling, Analysis and Testing of Autonomous Operation of an Inverter-Based Microgrid. *IEEE Trans. Power Electron.* **2007**, *22*, 613–625.

64. D'Arco, S.; Suul, J.A. Equivalence of Virtual Synchronous Machines and Frequency-Droops for Converter-Based MicroGrids. *IEEE Trans. Smart Grid* **2014**, *5*, 394–395.

65. Li, C.; Burgos, R.; Cvetkovic, I.; Boroyevich, D.; Mili, L.; Rodriguez, P. Analysis and design of virtual synchronous machine based STATCOM controller. In Proceedings of the IEEE 15th Workshop on Control and Modeling for Power Electronics (COMPEL), Santander, Spain, 22–25 June 2014; pp. 1–6.

66. Soni, N.; Doolla, S.; Chandorkar, M.C. Improvement of Transient Response in Microgrids Using Virtual Inertia. *IEEE Trans. Power Deliv.* **2013**, *28*, 1830–1838.

67. Hesse, R.; Turschner, D.; Beck, H.P. Micro grid stabilization using the Virtual Synchronous Machine (VISMA). In Proceedings of the International Conference on Renewable Energies and Power Quality, ICREPQ'09, Valencia, Spain, 15–17 April 2009; pp. 1–6.

68. Virtual Synchronous Machine. Available online: https://fenix.tecnico.ulisboa.pt/downloadFile/395145918861/paper.pdf (accessed on 17 June 2017).

69. Chen, Y.; Hesse, R.; Turschner, D.; Beck, H.P. Dynamic properties of the virtual synchronous machine (VISMA). In Proceedings of the International Conference on Renewable Energies and Power Quality, Las Palmas de Gran Canaria, Spain, 13–15 April 2011; pp. 1–5.

70. Chen, Y.; Hesse, R.; Turschner, D.; Beck, H.P. Comparison of methods for implementing virtual synchronous machine on inverters. In Proceedings of the International Conference on Renewable Energies and Power Quality, Santiago de Compostela, Spain, 28–30 March 2012; pp. 1–6.

71. Hirase, Y.; Abe, K.; Sugimoto, K.; Shindo, Y. A grid connected inverter with virtual synchronous generator model of algebraic type. *IEEE Trans. Power Energy* **2012**, *132*, 371–380.

72. Rodriguez, P.; Candela, I.; Luna, A. Control of PV generation systems using the synchronous power controller. In Proceedings of the IEEE Energy Conversion Congress and Exposition (ECCE), Denver, CO, USA, 15–19 September 2013; pp. 993–998.

73. Rodriguez, C.; Candela, G.; Rocabert, D.; Teodorescu, R. Virtual Controller of Electromechanical Characteristics for Static Power Converters. U.S. Patent US20140067138 A1, 27 February 2012.

74. Cortés, P.; Garcia, J.; Delgado, J.; Teodorescu, R. Virtual Admittance Controller Based on Static Power Converters. U.S. Patent US20140049233 A1, 20 February 2014.

75. D'Arco, S.; Suul, J.A.; Fosso, O.B. Control system tuning and stability analysis of Virtual Synchronous Machines. In Proceedings of the IEEE Energy Conversion Congress and Exposition (ECCE), Denver, CO, USA, 15–19 September 2013; pp. 2664–2671.

76. Zhang, W.; Remon, D.; Mir, A.; Luna, A.; Rocabert, J.; Candela, I.; Rodriguez, P. Comparison of different power loop controllers for synchronous power controlled grid-interactive converters. In Proceedings of the IEEE Energy Conversion Congress and Exposition (ECCE), Montreal, QC, Canada, 20–24 September 2015; pp. 3780–3787.

77. Behera, R.R.; Thakur, A.N. An overview of various grid synchronization techniques for single-phase grid integration of renewable distributed power generation systems. In Proceedings of the International Conference on Electrical, Electronics, and Optimization Techniques (ICEEOT), Chennai, India, 3–5 March 2016; pp. 2876–2880.

78. Johnson, B.B.; Dhople, S.V.; Hamadeh, A.O.; Krein, P.T. Synchronization of Parallel Single-Phase Inverters With Virtual Oscillator Control. *IEEE Trans. Power Electron.* **2014**, *29*, 6124–6138.

79. Johnson, B.B.; Dhople, S.V.; Cale, J.L.; Hamadeh, A.O.; Krein, P.T. Oscillator-Based Inverter Control for Islanded Three-Phase Microgrids. *IEEE J. Photovolt.* **2014**, *4*, 387–395.

80. Dhople, S.V.; Johnson, B.B.; Hamadeh, A.O. Virtual Oscillator Control for voltage source inverters. In Proceedings of the 51st Annual Allerton Conference on Communication, Control, and Computing (Allerton), Monticello, IL, USA, 2–4 October 2013; pp. 1359–1363.

81. Zhang, C.H.; Zhong, Q.C.; Meng, J.S.; Chen, X.; Huang, Q.; Chen, S.H.; Lv, Z.P. An improved synchronverter model and its dynamic behaviour comparison with synchronous generator. In Proceedings of the 2nd IET Renewable Power Generation Conference (RPG), Beijing, China, 9–11 September 2013; pp. 1–4.

82. Natarajan, V.; Weiss, G. Synchronverters with better stability due to virtual inductors, virtual capacitors and anti-windup. *IEEE Trans. Ind. Electron.* **2017**, *64*, 5994–6004.

83. Dong, S.; Chen, Y.C. Adjusting Synchronverter Dynamic Response Speed via Damping Correction Loop. *IEEE Trans. Energy Convers.* **2017**, *32*, 608–619.

84. Zhong, Q.C.; Konstantopoulos, G.C.; Ren, B.; Krstic, M. Improved Synchronverters with Bounded Frequency and Voltage for Smart Grid Integration. *IEEE Trans. Smart Grid* **2017**, doi:10.1109/TSG.2016.2565663.

85. Liu, J.; Miura, Y.; Bevrani, H.; Ise, T. Enhanced Virtual Synchronous Generator Control for Parallel Inverters in Microgrids. *IEEE Trans. Smart Grid* **2016**, doi:10.1109/TSG.2016.2521405.

86. Alipoor, J.; Miura, Y.; Ise, T. Stability Assessment and Optimization Methods for Microgrid with Multiple VSG Units. *IEEE Trans. Smart Grid* **2016**, doi:10.1109/TSG.2016.2592508.

87. Guo, W.; Liu, F.; Si, J.; Mei, S. Incorporating approximate dynamic programming-based parameter tuning into PD-type virtual inertia control of DFIGs. In Proceedings of the International Joint Conference on Neural Networks (IJCNN), Dallas, TX, USA, 4–9 August 2013; pp. 1–8.

88. Torres, M.; Lopes, L.A. An optimal virtual inertia controller to support frequency regulation in autonomous diesel power systems with high penetration of renewables. In Proceedings of the International Conference on Renewable Energies and Power Quality (ICREPQ 11), la Palmas de Gran Canaria, Spain, 13–15 April 2011; pp. 1–6.

89. Torres L., M.A.; Lopes, L.A.C.; Morán T., L.A.; Espinoza C., J.R. Self-Tuning Virtual Synchronous Machine: A Control Strategy for Energy Storage Systems to Support Dynamic Frequency Control. *IEEE Trans. Energy Convers.* **2014**, *29*, 833–840.

90. Shrestha, D.; Tamrakar, U.; Malla, N.; Ni, Z.; Tonkoski, R. Reduction of energy consumption of virtual synchronous machine using supplementary adaptive dynamic programming. In Proceedings of the IEEE International Conference on Electro Information Technology (EIT), Grand Forks, ND, USA, 19–21 May 2016; pp. 690–694.

91. Shrestha, D. Virtual Inertia Emulation to Improve Dynamic Frequency Stability of Low Inertia Microgrids. Master of Science thesis, South Dakota State University, Brookings, SD, USA, 2016.

92. Datta, M.; Ishikawa, H.; Naitoh, H.; Senjyu, T. Frequency control improvement in a PV-diesel hybrid power system with a virtual inertia controller. In Proceedings of the 7th IEEE Conference on Industrial Electronics and Applications (ICIEA), Singapore, 18–20 July 2012; pp. 1167–1172.

93. Chen, D.; Xu, Y.; Huang, A.Q. Integration of DC Microgrids as Virtual Synchronous Machines into the AC Grid. *IEEE Trans. Ind. Electron.* **2017**, doi:10.1109/TIE.2017.267462.

94. Bajracharya, L.; Awasthi, S.; Chalise, S.; Hansen, T.M.; Tonkoski, R. Economic analysis of a data center virtual power plant participating in demand response. In Proceedings of the IEEE Power and Energy Society General Meeting (PESGM), Boston, MA, USA, 17–21 July 2016; pp. 1–5.
95. Awasthi, S.R.; Chalise, S.; Tonkoski, R. Operation of datacenter as virtual power plant. In Proceedings of the IEEE Energy Conversion Congress and Exposition (ECCE), Montreal, QC, Canada, 20–24 September 2015; pp. 3422–3429.
96. Teng, F.; Strbac, G. Evaluation of synthetic inertia provision from wind plants. In Proceedings of the IEEE Power & Energy Society General Meeting, Denver, CO, USA, 26–30 July 2015; pp. 1–5.
97. Bousseau, P.; Belhomme, R.; Monnot, E.; Laverdure, N.; Boëda, D.; Roye, D.; Bacha, S. Contribution of wind farms to ancillary services. *Cigre* **2006**, *21*, 1–11.
98. Van Thong, V.; Driesen, J.; Belmans, R. Using Distributed Generation to Support and Provide Ancillary Services for the Power System. In Proceedings of the International Conference on Clean Electrical Power, Capri, Itlay, 21–23 May 2007; pp. 159–163.
99. Teninge, A.; Jecu, C.; Roye, D.; Bacha, S.; Duval, J.; Belhomme, R. Contribution to frequency control through wind turbine inertial energy storage. *IET Renew. Power Gen.* **2009**, *3*, 358–370.
100. Yingcheng, X.; Nengling, T. Review of contribution to frequency control through variable speed wind turbine. *Renew. Energy* **2011**, *36*, 1671–1677.
101. Yan, R.; Saha, T.K. Frequency response estimation method for high wind penetration considering wind turbine frequency support functions. *IET Renew. Power Gen.* **2015**, *9*, 775–782.
102. Can Synthetic Inertia from Wind Power Stabilize Grids? Available online: http://www.webcitation.org/6pscLoEBs (accessed on 20 April 2017).
103. Kempton, W.; Tomić, J. Vehicle-to-grid power fundamentals: Calculating capacity and net revenue. *J. Power Sources* **2005**, *144*, 268–279.
104. Almeida, P.R.; Soares, F.; Lopes, J.P. Electric vehicles contribution for frequency control with inertial emulation. *Electr. Power Syst. Res.* **2015**, *127*, 141–150.
105. Meng, J.; Mu, Y.; Wu, J.; Jia, H.; Dai, Q.; Yu, X. Dynamic frequency response from electric vehicles in the Great Britain power system. *J. Mod. Power Syst. Clean Energy* **2015**, *3*, 203–211.
106. Inoue, T.; Taniguchi, H.; Ikeguchi, Y.; Yoshida, K. Estimation of power system inertia constant and capacity of spinning-reserve support generators using measured frequency transients. *IEEE Trans. Power Syst.* **1997**, *12*, 136–143.
107. Zhang, Y.; Bank, J.; Wan, Y.H.; Muljadi, E.; Corbus, D. Synchrophasor Measurement-Based Wind Plant Inertia Estimation. In Proceedings of the IEEE Green Technologies Conference (GreenTech), Denver, CO, USA, 4–5 April 2013; pp. 494–499.
108. Ashton, P.M.; Saunders, C.S.; Taylor, G.A.; Carter, A.M.; Bradley, M.E. Inertia Estimation of the GB Power System Using Synchrophasor Measurements. *IEEE Trans. Power Syst.* **2015**, *30*, 701–709.
109. Lara-Jimenez, J.D.; Ramirez, J.M. Inertial frequency response estimation in a power system with high wind energy penetration. In Proceedings of the IEEE Eindhoven PowerTech, Eindhoven, The Netherlands, 29 June–2 July 2015; pp. 1–6.
110. Wall, P.; Regulski, P.; Rusidovic, Z.; Terzija, V. Inertia estimation using PMUs in a laboratory. In Proceedings of the IEEE Power & Energy Society Innovative Smart Grid Technologies (ISTG-Europe), Istanbul, Turkey, 12–15 October 2014; pp. 1–6.
111. D'Arco, S.; Suul, J.A.; Fosso, O.B. Small-signal modeling and parametric sensitivity of a virtual synchronous machine in islanded operation. *Int. J. Electr. Power Energy Syst.* **2015**, *72*, 3–15.
112. Golpîra, H.; Seifi, H.; Messina, A.R.; Haghifam, M.R. Maximum Penetration Level of Micro-Grids in Large-Scale Power Systems: Frequency Stability Viewpoint. *IEEE Trans. Power Syst.* **2016**, *31*, 5163–5171.
113. Agranat, O.; MacGill, I.; Bruce, A. Fast Frequency Markets under High Penetrations of Renewable Energy in the Australian National Electricity Market. In Proceedings of the Asia-Pacific Solar Research Conference, Queensland, Australia, 8–10 December 2015.
114. Gurung, A.; Chen, K.; Khan, R.; Abdulkarim, S.S.; Varnekar, G.; Pathak, R.; Naderi, R.; Qiao, Q. Highly Efficient Perovskite Solar Cell Photocharging of Lithium Ion Battery Using DC–DC Booster. *Adv. Energy Mater.* **2017**. doi:10.1002/aenm.201602105.

115. Xu, T.; Jang, W.; Overbye, T. An Economic Evaluation Tool of Inertia Services for Systems with Integrated Wind Power and Fast-Acting Storage Resources. In Proceedings of the 49th Hawaii International Conference on System Sciences (HICSS), Hostelling International, Koloa, HI, USA, 5–8 January 2016; pp. 2456–2465.

116. Garg, S.K.; Versteeg, S.; Buyya, R. A framework for ranking of cloud computing services. *Future Gen. Comput. Syst.* **2013**, *29*, 1012–1023.

117. Delille, G.; François, B.; Malarange, G. Dynamic frequency control support: A virtual inertia provided by distributed energy storage to isolated power systems. In Proceedings of the 2010 IEEE PES Innovative Smart Grid Technologies Conference Europe (ISGT Europe), Gothenberg, Sweden, 18 November 2010; pp. 1–8.

118. Baisden, A.C.; Emadi, A. ADVISOR-based model of a battery and an ultra-capacitor energy source for hybrid electric vehicles. *IEEE Trans. Veh. Technol.* **2004**, *53*, 199–205.

119. Pena-Alzola, R.; Campos-Gaona, D.; Ordonez, M. Control of flywheel energy storage systems as virtual synchronous machines for microgrids. In the Proceedings of the IEEE 16th Workshop on Control and Modeling for Power Electronics (COMPEL), Vancouver, BC, Canada, 12–15 July 2015; pp. 1–7.

120. Hao, H.; Lin, Y.; Kowli, A.S.; Barooah, P.; Meyn, S. Ancillary Service to the Grid Through Control of Fans in Commercial Building HVAC Systems. *IEEE Trans. Smart Grid* **2014**, *5*, 2066–2074.

121. Beil, I.; Hiskens, I.; Backhaus, S. Frequency Regulation From Commercial Building HVAC Demand Response. *Proc. IEEE* **2016**, *104*, 745–757.

122. Cao, Y.; Magerko, J.A.; Navidi, T.; Krein, P.T. Power Electronics Implementation of Dynamic Thermal Inertia to Offset Stochastic Solar Resources in Low-Energy Buildings. *IEEE J. Emerg. Sel. Top. Power Electron.* **2016**, *4*, 1430–1441.

MDPI AG

St. Alban-Anlage 66

4052 Basel, Switzerland

Tel. +41 61 683 77 34

Fax +41 61 302 89 18

http://www.mdpi.com

Applied Sciences Editorial Office

E-mail: applsci@mdpi.com

http://www.mdpi.com/journal/applsci

www.ingramcontent.com/pod-product-compliance
Lightning Source LLC
Chambersburg PA
CBHW051856210326
41597CB00033B/5916